Digital Transformation
Han

T0093453

Digital transformation has become more than a buzzword from the media since companies figured out the importance of rethinking business processes during global challenges.

On its own, the term assumes integration of digital technology into all areas of a business, resulting in fundamental changes to how the company operates and delivers value to customers.

Taking care of and choosing the optimal ICT tools is a constant struggle; the final decision may depend on the consultant's experience. Including all business stakeholders in this process is a must. Creating innovative company culture, continuous learning, and developing new skills with flexible and open communication and willingness to experiment are a challenge.

This complex, comprehensive approach can include implementing new systems, integrating existing systems, and using data analytics and artificial intelligence (AI) or machine learning to drive better outcomes.

By adapting and exploiting digital technology in new ways, businesses can gain better and more detailed customer experiences and build stronger relationships with their clients. In addition, digital transformation can help organizations to be more agile and responsive, which can lead to less time needed for different processes or the ability to adapt quickly to changing conditions.

In a time when change is the only constant – and it is hitting us every day not to forget about that – it is essential to think about digital transformation constantly. Technology improvement, availability, and scalability give us no room for excuses for not using them. Who can say that we are not living in dynamic and exciting times?

The Authors have taken their 20 years of practical experience and put it into this handbook, in which many cases can be found where not every time is a success story.

This book is prepared to provide some insights, give you a fresh overview of what such change can enable, and set up an environment for new technology that will arrive shortly.

Digital Transformation Handbook

Krunoslav Ris, Ph.D.
Milan Puvača, Ph.D.

CRC Press
Taylor & Francis Group
Boca Raton London New York

CRC Press is an imprint of the
Taylor & Francis Group, an **informa** business

First edition published 2024
by CRC Press
2385 NW Executive Center Drive, Suite 320, Boca Raton, FL 33431

and by CRC Press
4 Park Square, Milton Park, Abingdon, Oxon, OX14 4RN

ISBN: 978-1-032-30451-9 (hbk)
ISBN: 978-1-032-30452-6 (pbk)
ISBN: 978-1-003-30516-3 (ebk)

DOI: 10.1201/9781003305163

Typeset in Sabon
by SPi Technologies India Pvt Ltd (Straive)

Contents

**14 Next steps and what the future will bring: Instead of
conclusion**

Acknowledgments

Writing a book can be incredibly easy, especially when you have a support-ive family and a few close friends behind you every step of the way. This book is meant to serve as an acknowledgment and thanks to those who have stood by me throughout this process. I am so grateful to my family for their endless love and support during those chaotic times when I was between writing a book and day-to-day activities at Lumen Spei ltd.

Krunoslav Ris, Ph.D.

Dynamic environment seen globally in combination with IT as one of the most changing businesses are just two challenges faced during this book writing. Noting about processes which are constantly changing is always a dif-ficult task. However, I am sure we made a good knowledge base from empiri-cal point of view. Handling business, private and "writing" life is maybe even more complicated – but pulling it off is new personal satisfaction. My infinite thanking for all understanding as well as support is going to Ofir ltd. crew and of course my closest family - especially sons Andrej (5) & Filip (3). Hope that some parts of this book will be usable once you guys enter in world of work.

Milan Puvača, Ph.D.

Authors

Krunoslav Ris, Ph.D., is the author of the book *5G and Next-Gen Consumer Banking Services* (https://shorturl.at/osvFI). He is a senior digital transformation leader experienced in driving organizational change and managing high-growth agendas.

Kruno is a digital futurist who covers current digital transformation business processes. He has been an international FinTech speaker in over ten countries. Kruno has been in the IT business for 20 years and has worked on more than 400 projects. For the last 15 years, he has followed digital transformation processes worldwide on several projects across China, South Africa, Indonesia, Germany, Croatia, Canada, and USA. As the CEO of world-known Lumen Spei digital consultant agency, he was directly involved in the digital transformation of the most prominent Croatian National Institutes and Web3.0 Fintech Projects. His personal qualities, combined with diverse experiences in software development, mentoring, lecturing, software architecting, and strong leadership, give him a versatile perspective on upcoming technologies in the binary digital age.

This book is the result of spending more than 20 years bringing technology to life, leading digital strategy across multiple channels, transforming ideas that have led to multi-million-dollar businesses, and executing the profitable deployment of new technologies across multiple industries. He thrives on leading diverse teams and delivering operational agility in rapid growth environments with an entrepreneurial mindset. He is a mediator who builds bridges between Business, Operations, and IT. He is a capable organizer and motivator, and an experienced program/business process/IT manager.

Milan Puvača, Ph.D., is an IT solution specialist, consultant, professor, and mentor. He combines real-life experience with formal, structured education and development. He currently holds the position of CEO/ Owner of Ofir Ltd. (www.ofir.hr). He started his career in the IT sector in 2003, taking over the company which his mother founded in 1995.

From that point until today, Ofir Ltd. has been successfully growing, employing new people, and handling web development (designing, programming, implementing, and digital marketing) and computer/network maintenance services. In the academic field, he finished his Ph.D. studies in 2013 with the thesis "ICT as the foundation of the modern university" and is currently also engaged as a lecturer in College of Applied Sciences "Lavoslav Ružička" in Vukovar and faculty of humanities and social sciences in Osijek, Croatia.

He has gained experience by working with various people and handling numerous IT projects (network, hardware, web applications, and marketing strategies), and growing a company (financial, networking, and team management) pushes him towards further personal development as well as knowledge/skills dissemination. Milan has published more than 30 scientific and expert articles in peer-reviewed journals, conferences, and books. Through active membership in Osijek Software City and CISEx associations, he has prepared various lectures in the local community as well as Croatia and abroad.

This book presents his previous experience collected through uncountable working hours among different client profiles and business processes changes. One thing those people have in common are the idea of digital transformation and final mind/digital changing opportunity. Some of them succeeded, some didn't. Which group do you belong to?

About the book

In today's business world, the ability to be a truly agile company that can navigate the digitized world in which we all live and work is critical for competitive advantage, success, and business survival.

Researcher Marc Prensky 2001 coined the term "**digital native.**"

Digital natives have grown up under the ubiquitous influence of the internet and other modern information technologies. Digital natives think, learn, and understand the world around them differently from people who have not been as subjected to modern technology.

To describe people born after 1980 whose lives are shaped by access to networked digital technologies because they have never known any other way of life than the digitally mediated one.

Digital immigrants, on the other hand, use these technologies at a high level but have grown up in an analog world and are generally less familiar with the digital environment.

Similarly, digital natives have grown up in and been shaped by a digitally influenced world. Therefore, their worldview is not affected by old technologies, mindsets, cultures, strategies, or approaches. At the same time, some organizations have proven adept at adapting to the radically different environment in which they find themselves; this distinction is an important topic because nearly every area of every business (including customer interactions and expectations, operational efficiency and productivity, marketing and communications, sales, logistics, and distribution) has been significantly transformed by the impact of digital technologies.

Digital-native companies tend to have emerged from the technology sector but now span across various industries, from retail to logistics and marketing to automotive. What they have in common, however, is a natural, inherent ability to look at the world and the competitive markets in which they operate from a different perspective, to take an often contrasting approach to traditional problem-solving methods, and even to have a different "feel" for the values and organizational culture they embrace.

It's about skills, approaches, processes, and cultures shaped by our networked, technology-centric world, but it's also a mindset shift. It's as much about the behavior of individuals and team members as it is about the technology or digital solutions they provide.

Just as a person's first experiences shape them for the rest of their lives, companies moving into the digital world must also break down many outdated assumptions, ways of doing things, and organizational habits to transform themselves so that they are not only native speakers in the digital world but also doers.

We talk about digital transformation, not a digital adaptation because the change required impacts how things are done, how people work, how the business is structured, and how people feel when they walk through the door in the morning. In other words, it's about how a company works, behaves, and does business.

This book is about transforming businesses and business processes to serve their purpose in a digitally driven world. Our book seeks to capture, distill, and define the key lessons to help companies across various industries on their journey to transform into digital-native enterprises. In addition to our insights, we have woven into the text some "stories from the front lines" – contributions from other experienced digital transformation practitioners who bring to life their observations about how to do it right.

Chapter 1

What is digital transformation

Digital transformation is the process of utilizing digital technologies to either create new or alter existing business processes, culture, and overall customer experience to meet the changing business and market requirements. Also, it is a business reimaging process of surpassing all traditional roles of sales, marketing, and customer service.

Digital transformation is a holistic process that looks at the wider picture instead of focusing on particular business areas. By moving away from conventional business methods to a digital era, business owners have a valuable opportunity to reimagine how to do business with the immense potential of technology at their disposal.

Digital transformation allows both established companies and new businesses to strengthen their position in the market by digitalizing their business processes. More importantly, it is revolutionizing how entrepreneurs and business leaders think about business. To build a successful business in the 21st century, it will have to focus on thinking, planning, and building within digital environment (Figure 1.1).

Figure 1.1 Digital transformation cloud IoT.

DOI: 10.1201/9781003305163-1

SCHOLAR DEFINITION OF DIGITAL TRANSFORMATION

Digital transformation refers to the digital technology integration in all business areas intending to fundamentally change how a business operates and delivers value to customers. Also, it is a cultural change, requiring organizations to continue challenging the *status quo*, test and get comfortable with failure.

The world became increasingly digital, and businesses across all industries need to abandon outdated ways of doing business and start implementing digital tools, methods, and procedures wherever they can if they want to stay on top.

When seeking a unique, precise definition of digital transformation, the challenging part is its diverse implementation depending on the business needs. In almost every case, digital transformation will challenge businesses to walk away from their traditional business processes and embrace a range of new practices. However, such process can be completely different from one company to another.

Digital transformation puts the leadership and culture in the center, along with customer view, products and services, data, and all related technologies. For some companies, going digital will imply getting rid of the paper, while for others, it might lead to data analytics and artificial intelligence. Depending on the industry in which the company operates as well as internal company situation, digital will have a different meaning, yet its value will remain the same.

INDUSTRY DEFINITION OF DIGITAL TRANSFORMATION

The industry marks digital transformation as the third phase of the digital revolution in business. The first one was digitization, followed by digitalization, and lastly, there was digital transformation. In order to clarify potentially confusing terms, below are their brief explanations:

- **Digitization:** The process of moving from analog to digital in which businesses started converting all their paper records to digital computer files.
- **Digitalization:** The process of digital data usage to simplify and improve the way people work.
- **Digital transformation:** The process of adding value to each customer interaction by revisiting different business aspects to discover a more efficient way of delivering personalized customer experiences.

Simply put, the digital transformation process is reshaping the way businesses approach customer service. The conventional model was less proactive and dependent on customers to contact the company, whether in person or via phone. However, with the rise of social media, companies are being challenged to change how they advertise, market, sell, and service their customers. A more progressive approach allows companies to interact and build long-lasting, meaningful relationships with customers.

The goal of digital transformation is to convert companies into digital enterprises – organizations using all available technologies to continuously evolve their business models, including what they provide, how they interact with their customers, and how they operate.

Like evolution, digital transformation typically will not have a clearly defined endpoint. Businesses should evolve the same way technology evolves. They should no longer consider whether to transform their business or not, but how they will do it. By experimenting with different technologies and approaches, digital transformation can evolve these businesses and help them to rethink their status quo.

For enterprises, that means continually seeking ways to improve the end-user experience, whether through offering improved on-demand training, migrating data to cloud services, leveraging artificial intelligence, and more.

ACCELERATING TRANSFORMATION FOR A POST-COVID-19 WORLD

The COVID-19 crisis pressured companies across all industries to rethink everything about their digital transformation agendas. Companies needed to reinvent relationships with their audiences, including customers and employees. Most companies moved to remote work, another huge challenge that unquestionably affects all customer interactions.

In terms of digital transformation, COVID-19 has created several priorities for businesses like expanding the reach of customer support through interactive tools, automating procedures for resilience reasons, and an extreme housecleaning of redundant or conflicting systems. We can even talk about "forced digitalization" during COVID-19 situation.

As a response to the dramatic disruption brought by the pandemics, chief information officers (CIOs) started re-negotiating how these organizations connect with digital technology. That said, digital transformation is more of a transformation problem than a digital problem, which challenges the existing leadership to rethink their business in all areas (Figure 1.2).

CHALLENGES FOR BUSINESSES DURING THE COVID-19 PANDEMIC

One of the most evident challenges for businesses during the pandemic was to provide an alternative solution to the human touch. For years, we

Figure 1.2 Digital transformation COVID-19 world.

thought that customers sought in-person engagement to establish a relationship with a company, but COVID-19 proved us wrong. Due to a well-architected digital experience, customers have started revealing in a more personalized transaction from the comfort of their homes.

Apart from rethinking the nature of customer interaction, businesses in highly regulated industries needed to address barriers and provide solutions to ensure their customers can be offered the same quality through transformed services. One of them is the healthcare industry, where privacy concerns had to be solved first to allow healthcare institutions and individual healthcare professionals to provide telehealth consultations.

As the world slowed down, businesses reduced their pace and concentrated on becoming unique instead of following the new trends. Since being passive on the market rarely produced the desired results, companies had the perfect opportunity to upgrade own processes by analyzing the competition with emphasizing their strengths. Organizations that started changing their mindset are at lower risk for business closure than businesses still following others in the industry.

Although the lead digital transformation player, the IT sector, often didn't meet all the requirements to keep up with digital transformation efforts; for IT departments in numerous companies, the pandemic brought out their maximum to ensure operations, and revenue have not been affected negatively.

Lastly, another incorrect assumption was that people will refuse to pay the full price for digital formats. The recent events showed that consumers had no problem paying for digital products and services. On the contrary, it proved to facilitate the entire customer experience.

FACILITATING DIGITAL TRANSFORMATION ONWARDS

As both organizations and customers are gradually embracing digital transformation, companies need to leverage this shift and simplify the transformation process in the future. The focus of these companies and their next-generation digital initiatives should mainly be on customer experience, employee experience, operations, and business model transformation.

Customer experience is becoming more personal and requires more emotional involvement than before. With access to data, companies are leveraging the information about their customers to ensure a more personalized and valuable experience, using the most recently developed technologies in the field of artificial intelligence and machine learning.

These two technologies can also increase efficiency when completing routine tasks, along with the augmented reality that assists employees in ways not possible before. Simplifying and improving employee experience allow companies to re-establish the relationship with their employees, strengthen the company culture, and attract new talents that match company values.

The Internet of Things and Industry 4.0 allow companies to improve their operational performance and introduce new services. Some companies have already introduced innovations such as machine learning and digital twins that brought more value from real-time data.

As much as digital transformation is the way of the future, companies can also seek out smaller opportunities that will lead them to digital improvement and information-based extensions. For example, the companies in the insurance sector are monitoring and scoring their customers with the ultimate objective to optimize policy pricing.

IMPACT OF THE DIGITAL

Researchers studying scientific progress and technical change in today's society say that digital transformation comes from general-purpose technology. This means that such technology has the potential to evolve itself, continuously branching out and enhancing productivity across all industries. The general-purpose technology will bring immense long-term benefits, but its highly disruptive nature is challenging to implement (Figure 1.3).

That said, most benefits come from not only adopting the technology but also from adapting to it. In modern society, one of the best examples is Uber, a taxi company that utilizes digital technology to provide a more quality service to its customers. Before Uber, there was no other platform to search city rides, leave reviews, and interact with the drivers.

Before society adapts to any disruptive technology, it must be widely adopted first. The technological revolution depends on computers, the Internet, search engines, and numerous digital platforms.

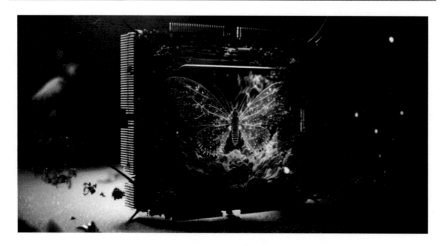

Figure 1.3 Digital impact.

Adapting to new processes takes time before it starts demonstrating its value in replacing its outdated alternative. Innovation and reorganization were always the first pillars of any revolution throughout history.

Although he marketed quite an efficient engine in 1774, James Watt had to wait until 1812 to see the first commercial steam locomotive. It took almost 40 years to demonstrate its benefits for society, so there is no surprise in why there is still so much vagueness around personal computers that appeared 40 years ago.

That, however, doesn't imply that the impact of digital is little. Besides transforming jobs and skills, digital transformation also overhauls numerous industries such as publishing and retail. Solely in the United Kingdom, Internet transactions already report for 20% of retail sales. Compared to 5% from 2008, it is safe to assume that this percentage will continue to grow steadily in the future.

It makes no sense to talk about digital revolutions without mentioning the leading technological breakthroughs of digital – blockchain technology and cryptocurrencies. Blockchain technology is not yet another solution for online payments. It is also an efficient solution against any illicit activity like money laundering.

The objective of this underlying technology is to fully revolutionize finance by making all transactions more secure and faster while having access to quality information on potential customers can even enhance loan pricing through better analysis of the chances of repayment. In this situation, regulatory institutions are crucial to provide financial integrity and protections to the customers while also focusing on efficiency and innovation.

When looking at what is coming, we can expect to see more disruptions from breakthroughs in technology like quantum computing, which aims to simplify calculations that exceed the capabilities of conventional computers.

A few benefits will become evident once the technology is applied. For instance, these computers can potentially undo some of the new technologies like rendering existing standards in cryptology obsolete, which might affect communication and privacy worldwide.

ACCELERATED PACE

One of the immense impacts digital has on our society is the job transformation. Experts predict that almost one-third of the workforce in the United States might transform in the following years. Also, up to 50% of all paid activities could become automated due to technology like artificial intelligence and machine learning.

Today's computers are learning how to drive public transportation and check for signs of diseases, which was previously done only by specialized doctors. In many industries, jobs will be lost or transformed, including those considered not to be affected by automation.

Regardless, the pace of the transformation around us will continue to accelerate. When looking at one piece of technology – smartphones, we see more than 4 billion people accessing enormous computing power. So, the question is not whether someone approves the spreading of digital technology or not, but how can we make the most of it. If anything, history has taught us that humanity will adapt to using steam power and electricity, and chances are that the same will happen with digital transformation.

COOPERATION INSTEAD OF A RACE

The global reach of digital technology is undeniable, so avoiding a race to the bottom and focusing on policy cooperation in global financial markets is vital. This cooperation should regulate personal data management, surpassing the country borders because, after all, the Internet is more international than national (Figure 1.4).

This cooperation also requires global organizations like the International Monetary Fund and the World Bank to be placed in the process. These institutions have everything necessary to provide a forum for addressing challenges brought by the digital revolution, work on efficient policy solutions, and start paving the way for revised policy guidelines. The value of reorganizing the economy around these revolutionary technologies will undoubtedly result in immense long-term benefits.

Going back to the job market, the created gig economy requests rule reconsideration in many aspects. One of them is the lack of clarity on the meaning of self-employment where individuals have more flexibility when earning money in companies like the mentioned Uber. The need to reduce disruptions and increase benefits requires revisiting policies around

Figure 1.4 **Digital race.**

international taxation and digital data, labor policies, and inequality to this digital revolution. Having quality policies in place and the needed willingness to cooperate beyond borders and any other conventional limitations will allow us to harness these emerging technologies to enhance wellbeing without excluding what the digital age brings to society.

PREDICTIONS ON DIGITAL INNOVATION

One of the frequent digital transformation predictions is the connection strengthening between people and the devices used daily. In the following years, that connectivity will continue to grow as more digital applications appear and provide people with an opportunity to create, share and participate.

This trend will probably accelerate with the growth of the employment of smart agents and bots for interaction with others. This continuous and spreading human connectivity impacts how we engage as citizens and potentially changes how we interact with democratic institutions.

Regarding devices, estimation is that the number of connected devices, such as wearables, cars, household appliances, will continue to grow. People are witnessing great benefits from connecting on a deeper level with their environments. This connectivity will continue growing in connected devices and expand across new areas. The objective is to create even more connected environments, such as smart buildings, smart streets, smart plots of land, etc.

That will lead to advancing the level of knowledge people possess about themselves and everything around them, which could drive policy change. Another aspect of the digital era will continue evolving over the next few years – data. The data explosion will impact social and civic innovation in a few possible directions. The most plausible direction is that privacy issues will have a strong effect on norms and behaviors. Also, how this immense volume of data is analyzed will bring more critical observation around the algorithm performance and whether the way that data is being used is fair and explainable.

DIGITAL TRANSFORMATION – WHAT DOES IT LOOK LIKE?

The digital transformation process needs to be implemented in three key areas: customer experience, business models, and operational processes. Every organization must aim to understand their customers better, and with technology, there are numerous opportunities to fuel customer growth and create new customer touchpoints.

Also, digital transformation requires organizations to enhance their internal processes by leveraging digitization and automation while also providing their employees with digital tools, gathering data to monitor performance, and making better-informed business decisions.

When discussing business models, organizations will transform their businesses by increasing physical offerings with the digital format. They will also introduce digital-based products and use technology to offer global shared services.

DIGITAL TRANSFORMATION EXAMPLES

We are seeing digital transformation happening across almost every industry and job function. To have a better understanding of how digital transformation works, it is helpful to look at the examples.

In a sales department of a typical organization, digital transformation might involve migrating spreadsheets to a cloud CRM. An expected result of such a process is improving win rates, streamlining customer relationships, and enhancing customer data by implementing a CRM solution.

To digitally transform an HR department, in-person training might be replaced with online learning by using online audio or video communication tools. This shift can help with onboarding quality, overall training costs, and automate their other HR processes.

In customer support, a digital transformation process might involve replacing a conventional call center with an online knowledge base.

When it comes to industries, digital transformation will differ depending on the problems it aims to solve. For instance, in the healthcare industry, digital transformation can lead to virtual visits, telemedicine, and patient portals, whereas in the hospitality industry, it might involve online check-in and amenity booking tools. Insurance companies might implement virtual quotes and online claims processes as a result of digital transformation, and retail companies might introduce loyalty cards or e-commerce stores.

CRITICAL AREAS FOR A SUCCESSFUL DIGITAL TRANSFORMATION

For any organization implementing digital transformation, regardless of the industry, their CIOs will need to focus on these areas to achieve success:

Implementation of digital twin

Also known as the digital representation of an entity or system in the real world, the digital twin implementation involves a software model that mirrors an individual physical object, person, process, or organization. Their purpose is to support the entire process of digital transformation because they simplify experimentation and gather data that supports better-informed decisions.

Privacy

Without efficiently managing privacy, digital transformation cannot be successful. With an increasing number of digital solutions available to organizations, convenience has become one of the deciding factors. However, consumers and employees are not willing to compensate for their privacy and safety for convenience. That is why CIOs must build everything around privacy or lose the support of their customers and employees, which would affect their business in unimaginable ways.

Culture

It is a part of human nature to resist change, although it leads to a positive outcome. Ignoring the cultural consequences of a digital transformation will result in resistance from your employees and customers. Many CIOs report that culture is the biggest obstacle to any change they want to implement in their business, so addressing culture is essential for a successful digital transformation. One of the ways to introduce change to your employees is by having change leaders who vocally support your digital transformation and impact how others feel about it.

Augmented intelligence

Augmented intelligence is a wider term than artificial intelligence (AI), permitting people and machines to work together. The data collected and analyzed with the help of AI is more valuable than anything a human can deliver. Yet, augmented intelligence doesn't aim to replace employees with machines but only gather and collect data. Then, people can augment their existing knowledge and be of more value to their organizations.

Digital product management

Digital product management involves shifting of mindset from one project to another. Such products have to be designed to improve the overall customer experience and deliver through any digital channels. In digital product management, both industry knowledge and product design are required. A great example of digital product management is when Apple created watches that monitor the health of the person wearing them instead of waiting for the entire healthcare industry to align with the company offerings.

WHAT ARE DIGITAL TRANSFORMATION DRIVERS?

When looking into digital transformation as we see it today, we can notice five key drivers over the past few years. These drivers are contributing to digital transformation in their unique ways. They are also bringing specific results that continue driving the change (Figure 1.5).

Figure 1.5 **Customer experience.**

CUSTOMER EXPECTATIONS CONTINUE GROWING

Compared to only a decade ago, today's customers are much more empowered than they used to be. They are becoming increasingly savvy and expect a truly personalized experience from businesses and organizations. More importantly, this personalized experience needs to be consistent across all used communication channels, from the website and social media to direct messaging tools.

Today, customers prefer to spend their money with brands that understand the importance of a personalized shopping experience and will get frustrated when their identity and action have been loo when switched to another communication channel.

The reason behind these growing expectations lies in companies that dared to be the first to create customer-centric platforms, such as Amazon, Uber, and Netflix. Every time a customer interacts with one brand more engagingly, their expectations will immediately grow and be applied to other businesses as well.

A more radical view of these digital transformation drivers names today's customers "digital predators". Because they have the power now, customers can change the markets as we know them by leveraging the power of technology to raise customer expectations to an even higher level. For companies, this means that digital transformation must be a high priority if they wish to avoid becoming "digital prey".

SPEED IS CRUCIAL

One of the customer expectations that need to be improved is the speed. Companies must prioritize becoming faster in interacting, serving, delivering, and re-engaging their customers because the digital race can only be won by the fast ones, not the big ones like it used to be the case before.

The biggest obstacle to velocity is the current mindset the companies have. To be faster across all phases of the customer journey, businesses have to change their linear approach. Digital transformation is not linear, and providing solutions with a linear mindset will not bring the desired results. Once they have determined their perspective or the way to respond to an opportunity, things will change again, and they will need to restart the process. The major challenge is learning how to move rapidly towards the future without stopping when any digital disruptors appear.

STRONG FOCUS ON CUSTOMER INSIGHT

The ability to rapidly respond to all customer expectations will always start with a firm focus on customer insight. Those companies that are considered successful are inventing and designing demonstrate how deeply they understand their customers and their needs. To develop this intuition, they will spend an immense amount of energy on studying and understanding the relevant aspects of the lives of their customers rather than the basic information they get from superficial surveys.

Customer insight needs implementation at several levels, from understanding the entire customer base to understanding the relevant needs of each customer to ensure they get a personalized experience when interacting with your company. However, achieving that level of insight is not that easy. Companies need to develop customer intuition and gather anecdotes, but also leverage data to continue being aware of their customers' preferences.

EVERYTHING REVOLVES AROUND AI

Undoubtedly, artificial intelligence (AI) is one of the fastest emerging priorities out there. Siri and Alexa, the pioneers of the conversational interface based on AI, set an example for other companies to follow when implementing a strong AI strategy to respond to the demands of their customers. Yet, the meaning and the potential of AI still stays unclear to many.

What value AI provides and how it can be applied will significantly depend on the industry and nature of the business itself. There are already numerous Ai interaction models that serve as an inspiration to those considering implementing this technology. Regardless of how a company decides to use AI to introduce improved customer experience, the successful implementation must be sensing, comprehensive, guided, actioned, and with great analytical features.

DIGITAL TRANSFORMATION EQUALS BUSINESS TRANSFORMATION

Most digital transformations lack the business perspective when evolving because the true digital value proposition goes beyond the digital. To be successful, it needs to reinvent numerous aspects of the business, often including core technology systems, business models, and vital operating processes. Those who see digital transformation as the fundamental shift in the way a company delivers values and drives revenues are the industry leaders of tomorrow.

Digital transformation comes with consequences, and sometimes, the positive ones will outshine the ones that require additional effort. For instance, when Hilton created a mobile app that allows their customers to unlock doors on their own, besides facing a software development challenge, the company had to change dozens of thousands of locks at their properties and retrain the team on the repairment and maintaining of the new locks.

OTHER TRENDS DRIVING DIGITAL TRANSFORMATION

Although these are five key digital transformation drivers, other significant trends challenge companies to adapt to the new world around them. One of them is social media and mobile technologies. The devices we are using, smartphones and tablets, are changing how we interact and behave. This change leads to a long and exhaustive process of adapting to the dynamic needs of both employees and customers and finding a way to offer relevant opportunities through digital tools.

The 24/7 concept is also something that has become a quality standard across industries. One of the growing expectations was also for companies to be reachable at all times. There is no more waiting for days or even hours to get an answer. The more customers need to wait, the higher the chances are of abandoning the brand. Going back to the velocity, being fast will also include the individual interaction with the customer and ensuring they get everything they need, from information to product, as quickly as possible.

Besides trends outside the company, changes are happening within the organizations too. To constantly deliver new solutions while also being faster than others will require teams to work differently. A few years ago, departments were separated by fields, and each team was working on one product area. Now, all these teams have to work together on satisfying empowered customers. Only that will allow them to exchange information more accurately and quicker between them, and ensure they are keeping up with the market demands.

One term connected to digitization, especially when talking about companies that produce outside their halls is "extended enterprise". It refers to the ability to exchange the content and processes with all involving partners with the help of digitalization. A product designed in Italy can be developed by engineers in Germany and produced in China. All these constructs are a part of the so-called "extended enterprise". This is one of the determining digital transformation drivers for manufacturers, but it can also be applied to any other company, especially now with everyone moving to digital.

Undoubtedly, AI has an immense value in the digital transformation process, but it is not the only new technology we know. With data becoming more important, companies are focusing their efforts on it. Big data is already revolutionizing how we process unstructured information such as videos, photos, and emails, which will save a lot of time for those companies using their resources to search for valuable information.

Along with the immense amount of data, companies are also exploring the potential behind the Internet of Things (IoT). Such technology adds more intelligence to the existing machines, yet they still need to be monitored. The objective is to optimize business processes, develop new products, and manage and monitor in general.

Lastly, companies can now utilize hybrid cloud IT solutions and infrastructure instead of relying only on their servers. These cloud solutions lead to reducing IT infrastructure costs. Also, they save the time it takes to install, upgrade, or make any other changes to the existing server. Without a doubt, cloud IT solutions are the only way because the companies are also becoming dependent on them due to the high degree of mobile users who expect every aspect to get done online.

TYPES OF DIGITAL TRANSFORMATION

Any organization considering going through digital transformation, regardless of its industry, will need to consider four main areas of it: process transformation, business model transformation, domain transformation, and cultural or organizational transformation.

PROCESS TRANSFORMATION

Process transformation includes adjusting the elements of processes within the company to achieve new objectives. Companies will typically engage in a process transformation when needing a radical update. Such transformation aims to modernize the business processes, implement new technologies, save financial resources, and more.

Business process transformation is similar to business process management, but the goal is to drastically improve the way business works. These are the main steps an organization will need to take if it wants to transform its processes:

1. **Determine the goal of the transformation**: Is the company looking to upgrade systems, incorporate new technologies, or adapt processes to their new organizational structure?
2. **Decide on baseline metrics**: Companies need to gather the data, such as costs and time, to prove that the process transformation leads to success.
3. **Gathering all stakeholders**: Approach everyone involved and ask for feedback on the previous and new processes.
4. **Craft the best scenario**: Use a diagramming tool to create the desired workflow path, requiring human and system tasks.
5. **Launch and track**: Introduce the new process to small teams first, and monitor the progress and needed changes to ensure success.

BUSINESS MODEL TRANSFORMATION

Many companies pursue digital technologies because they want to transform conventional business models that no longer work or produce the results they used to. Almost across all industries, there are examples of such innovation. Netflix is reinventing video distribution and Uber the taxi industry. These are the leaders in their industry, but there are many more examples of smaller companies incorporating these significant business model changes. Companies in the insurance sector are using data and analytics to charge separately by the mile instead of traditional insurance contracts.

By rethinking and changing their current models to achieve more success, companies can find new opportunities that lead to growth. That is why the transformation of the business model will be a priority for companies wishing to stay ahead of their competition.

DOMAIN TRANSFORMATION

Unlike process and business model transformation, domain transformation is not so much talked about, although it provides enormous value for organizations. With new technologies, companies are redefining their products and services, blurring industry boundaries, and fostering opportunities for non-conventional competitors. This type of extensive transformation provides valuable opportunities for companies.

A great example of how the domain transformation enhanced an entire business is Amazon, the online retailer. The company expanded into a new market domain when they launched their Amazon Web Services or AWS. At the moment, AWS is the leader among cloud computing/infrastructure service providers, leaving behind corporate tech giants like Microsoft and IBM.

To start providing cloud services, Amazon only had to leverage the capabilities and services the company was already using, accelerating the process tremendously when moving into the new space. Those companies that decide to undergo a digital transformation need to be aware of the potential of domain transformation when implementing new technology.

CULTURAL OR ORGANIZATIONAL TRANSFORMATION

It is not enough to update your existing technology or redesign products for a successful digital transformation. Often, businesses will fail to align these digital transformation measures with their internal values, which leaves a heavy effect on the business culture. Negative consequences will range from slow, complicated adoption of digital technologies to a complete market competitiveness defeat. This leads to revenue and productivity drop, which can be the end of an organization.

When you implement the right one, cultural or organizational transformation can help shift the entire business culture to understand, embrace, and extend digital transformation. Leaders of these organizations need to create a clear transformation vision and communicate it successfully throughout the entire organization. Besides that, they need to be aware of the smart risks in front of them and why they are worth taking.

When it comes to digital culture, these are the core elements of it:

- Promoting external instead of internal orientation
- Appreciates delegation over control
- Focuses more on actions and less on planning
- Aims more for collaboration instead of individual effort

THOSE WHO CLOSE THEIR EYES TO DIGITALIZATION RISK FAILURE AND EXTINCTION

Although no one expects managers to have clairvoyant abilities, the example of the former global brand Kodak shows what happens when corporate management refuses to embrace digital transformation.

Blessed with creative developers, Kodak Labs presented the world's first digital camera as early as 1975. But management put the brakes on the project, fearing that the novelty would hurt Kodak's highly profitable film business. Instead, competitors from Japan did so in the 1980s.

When Kodak finally started making digital cameras, it was too late, and the early advantage was lost. By 2012, Kodak was bankrupt and its $35 billion market value was gone.

New technologies spread faster than ever to first 50 million users (Figure 1.6).

In the meantime, even the market for digital cameras has become a niche market, but who knows, if Kodak had dared to take the step into the digital age in 1975, a learning curve like Apple's might have been possible. Maybe the first iPhone would have been built by camera manufacturer Kodak and not by computer manufacturer Apple.

IF THINGS ARE GOING WELL, WE NEED TO CHANGE

Companies that do good business have a tough time suddenly reinventing themselves to ensure tomorrow's sales and profits. Early indicators of change should be considered more critically. Even today, we are experiencing a digital revolution like Blockbuster, Inc. In 2004, Blockbuster was the largest video rental company in the United States, with 8,000 stores and $6 billion in sales.

No one on the board of the powerful market leader took seriously Netflix; the competitor founded just a few years earlier that allowed customers to rent DVDs online and receive them by mail, with attractive subscription models.

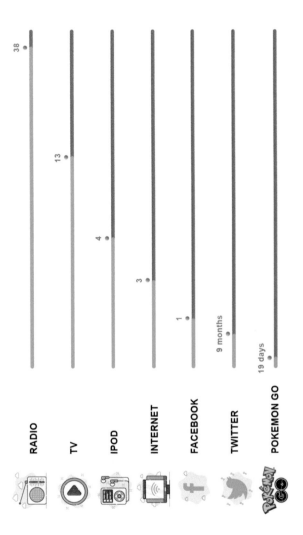

Figure 1.6 Impact of the media.   Source: McKinsey Global institute.

Blockbuster engineers worked on a system for online ordering without any sense of urgency. In 2007, Netflix made a giant leap forward by offering video on demand – movies that could be streamed directly over the Internet. DVDs were obsolete. Customers flocked to Netflix because the offer was attractive:

- No waiting for mail delivery
- No returning to the post office
- Instant enjoyment

Only then did Blockbuster react and develop its video-on-demand system, but it needed to be better and of better quality. Netflix had already captured a significant market share, and Blockbuster offered no innovative new features to win back customers – and to make matters worse, Blockbuster's service and delivery were worse than Netflix's. Netflix had immediately won over the Internet-savvy, mostly young customer base.

And in just a few years, the vast majority of movie fans discovered how easy it was to spend an enjoyable evening with Netflix. Today, Netflix is the market leader, while Blockbuster filed for bankruptcy in 2010 (Figure 1.7).

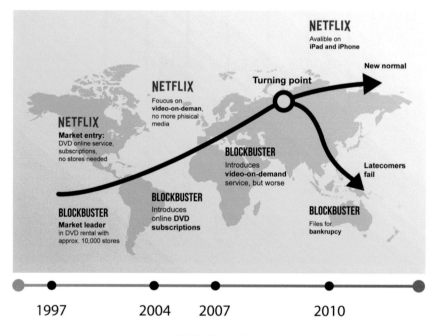

Figure 1.7 Netflix – Blockbuster.

Replaced DVD rental with a digital online offering – despite all its efforts, market leader Blockbuster was unable to survive

The lesson from this example is clear: regardless of how well-positioned a company is if the management underestimates the potential for change that digitization poses to its business model, it runs the ultimate risk. And those who see the change but delay their response not to jeopardize their current revenues are taking a virtually suicidal stance.

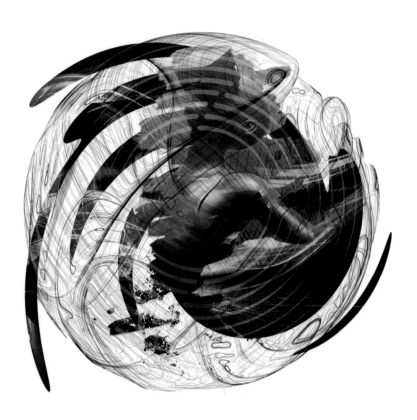

Chapter 2

Digital transformation evolution

If we look at the nature of digital transformation from its first moment to today, it's impossible to ignore its evolution. The meaning of digital transformation is also continuously evolving. Firstly, it implied turning paper into digital form. Then, we used this term to explain process optimization by using software solutions. Naturally, what followed was networking, software as a service (SaaS); and lastly, we have a new digital economy.

Undoubtedly, the pandemic accelerated digital transformation and changed its meaning once again to explain the change from optimizing a business to transforming it. Climate change, social unrest, trade wars, and automation processes in factories were the main external factors of this change. Adding the pandemic to all of these factors, digital investments were accelerated even further. As a result, we are now having digital transformation that was expected to happen in the next 10 years or so (Figure 2.1).

A simple example that demonstrates yet another change of the meaning of digital transformation, which can also be seen as the most recent revolution of it, is through a business organization. The emphasis is no longer on revolutionizing the organization's business with digital advancements, but on the evolution. Here, digital evolution refers to harnessing all digitalization potentials to improve things as they are now.

Digital evolution allows companies to use what they have instead of throwing it away and building something completely different. After all, organizations spend years building their procedures, and doing everything from scratch doesn't sound like optimization. The evolution in digital transformation is built around incremental advancements that can improve businesses in any industry with gradual changes, which are the pillars of digital evolution. When approached this way, digital transformation evolution is an efficient and necessary solution for both big and small organizations.

DOI: 10.1201/9781003305163-2

Figure 2.1 Digital transformation evolution.

LETTING GO OF THE PAST

Only a few years ago, digital transformation was the process of data modernization, core technology improvements, and back-office automation, yet today's challenges differ significantly. Each organization is seeking ways to digitize its end-to-end operations, but that's not the ultimate goal of digital transformation. The digitization of these processes should lead to simplifying how organizations are adapting to constantly changing market dynamics.

The work that is required to digitally transform an organization is more complex because companies need to shift from optimizing to reinventing and completely disrupting not just the lives of their customers but their employees as well. Simply put, to have products that disrupt and enhance the lives of customers, companies also need to disrupt the way their employees work.

The potential of current technology to transform almost every aspect of the company is an excellent opportunity to reinvent the way to do business and also expand their reach. On the other hand, as the game is becoming more complex, the competitors are also difficult to compete against. In each industry, most businesses are already embracing these new tech-supported business models and finding ways to reinvent themselves; therefore, being unique and on top of the game is a bigger challenge than ever before. Finding the value in the way a company does business is becoming as important as the quality of its products or services.

Digital transformation evolution implies almost the same amount of challenges as opportunities. Will supply chains find a way to become more agile in scaling up and down? Are remote employees equally productive and collaborative as when in in-person teams? How can businesses anticipate and respond to customer demand? To these and many more rising questions, companies will need to seek their answers in a different way than they used to.

Instead of having *ad hoc* solutions that are putting out fires, organizations should aim for subtle, yet more complicated transformations that require additional resources when compared to the previous ones. How a company manages its transformations will determine its overall value, brand perception, customer relationships, sales, and more.

That said, most businesses still don't have the right digital basis in place as their biggest struggle is to link their digital investments to the company's strategic value. When compared to the previous transformations, this one has stronger and more visible consequences. When an organization is focused on investments in their analytics to improve their finance group's performance, it doesn't affect how the external audiences see them. If the project fails, the company's customers wouldn't even be aware of it.

Now, on the other hand, any inefficiency will be less tolerated as the stakes are high. It wouldn't be unusual to see a business shut down due to a digital transformation gone wrong. Seizing an opportunity doesn't just help an organization in a race with its competitors but also prevents a catastrophe like shutting down your supply chain, destroying your recently launched product, or putting the business to sleep for good. More than ever before, it's crucial to get the digital transformation right.

The success of an organization's digital transformation lies in putting strategy first, even before technology. As much as digital transformation revolves around all these technologies, it will not produce the desired results if it's not aligned with a carefully thought strategy. All organizations that wish to survive all challenges put in front of them due to the mentioned external factors, and there will be even more of them, should first determine what their business needs and then seek a technology that can provide the best solution to those needs.

A successful digital transformation will require an honest conversation and effective collaboration between business and technology leaders. Any miscommunication or lack of it can stall the progress of transformation and decrease the final output. More importantly, without these two roles working together, a company's digital transformation will be all about technology and strategy will fail.

Regardless of the industry, each business will first need to reply to the question "why" to get to the "what", meaning they need to know why they are implementing a certain technology, and not just what technology will be implemented.

Not to say that just implementing technology will lead to failure. Such an approach has short-term benefits for the business; however, they will soon need to be replaced with new solutions. This lack of long-term strategy might work for several months or even years, but ultimately, it will lead to bad results, both internally and externally.

The understanding between business and technology is essential with digital transformations. If their incentives aren't aligned, companies will typically abandon the decision-making process and leave it up to the IT department. However, without the business and technology team working together, the conversation will not follow the business strategy which factors in customer insights necessary for the company to succeed.

FINDING COMMON GROUND

Addressing this lack of collaboration is one thing, but finding efficient solutions for business and technology leaders to work together seamlessly is another. So, how can both teams truly understand each other and find common ground? They both need each other, as the business cannot discuss technology without the tech team, and the tech team cannot understand the strategy without the business team.

To ensure a company's efforts towards digital transformation are producing results, forgetting about technology at the beginning of the process is necessary. To avoid miscommunication and maintain the right focus, organizations should focus on these five imperatives to build the right digital foundation:

1. **Experiences**. Both employees and customers want better experiences and to succeed as a business, companies need to reinvent how they provide value to these two audiences.
2. **Insights**. For an informed, valuable decision-making process, companies need new and improved insights.
3. **Platforms**. Information is everything, and finding the right platforms to deliver that information is crucial.
4. **Connectivity**. Chosen platforms allow companies to get their information to the right people with ease and much faster.
5. **Integrity**. Companies need adaptive digital infrastructure that stands against any bad factors which might affect how your audiences perceive your company.

These five imperatives of digital transformation need to be communicated to the company's employees. Going one step further, companies that take their digital transformation process seriously should have a conversation with their employees about these fundamentals to hear their opinion and reach a consensus among everyone on what each of these five imperatives means for the business. Besides clarity on the meaning behind each of the terms, business leaders will also avoid miscommunication and making poor decisions during the process of digital transformation (Figure 2.2).

For instance, if a team manages to solve a data problem that is blocking the digital transformation process in the company, because of the conversation everyone in the company had, each stakeholder will understand that the team didn't just clean up data, but they also improved overall customer experience, insights, and agility. These five pillars of digital transformation allow technology teams to adopt a business-outcome mindset, while the business team learns to communicate in a way that makes sense for IT. That said, they can do much more than just help two or more teams agree on common goals and strategies.

Being clear on what each term represents for the company helps promote empathy and limits confusion among employees, which greatly

Figure 2.2 Digital city.

reduced the possibility for any negative outcome. Also, these digital transformation imperatives promote adoption because all employees understand what needs to get done and the reason behind it, so their communication is clear, time-efficient, and productive.

Without a doubt, the tech will continue to evolve over the next few years as well, but at a much faster rate. By seeing conditions producing a need for high stakes, it's expected to only intensify. Those companies that focus on building all of their efforts around strategy, and not technology, will be able to build a plan that will deliver value over a longer period, regardless of the type of tools these companies choose to use. A good strategy will not be relevant only for a few months, but for a few years. If focused on technology, companies are setting themselves to revise and rebuild their strategies every few months because technology changes, and the only thing that stays strong even with all these changes is a quality business strategy.

We're living in an era of the biggest disruption so far. We have learned what potential lies in all these digital technologies around us, and businesses are making the leap to a new, reimagine future to cross over to from the current crisis. That's why companies need to be bold when working on their vision and the ways they want to transform their business in the future, but be constantly aware of all consequences surrounding them. And, most negative consequences for businesses in this digital era start with not being on the same page with your team.

DIGITAL TRANSFORMATION EVOLUTION BENEFITS

As digital transformation is evolving, there will be new benefits organizations across all industries are noticing. Up to now, there have been several long-term benefits that are motivating not only those who are at the beginning of their digital transformation process but also those who are still not aware of the immense potential this transformation can bring them.

Speed

Companies that have taken a digital evolution approach will be able to get things done faster. When compared to previous performance, businesses marginally reduced the time required to develop and deploy an idea. In business, time is often money, so spending time on processes that might be optimized, automatized, or enhanced in any other way during a digital transformation process (Figure 2.3).

Affordable

Many wrongly assume that because of its big impact, digital transformation will cost organizations a lot of money. On the contrary, because it's focusing on the current development, digital transformation will lower the company's

Figure 2.3 **A vision.**

costs and, at the same time, increase ROI. When businesses decide to create something from the scratch is expensive as it requires plenty of research, testing, new technologies, and skills, and there is always the possibility of things not turning as you hoped and investing again your resources into fixing it.

Low risk

A transformation will not happen overnight. Whichever changes a business is implementing as a part of their digital transformation process, it will happen gradually. This eliminates the risk most companies are afraid of, and that's jumping into an unknown and complex area that is eliminating everything they've built from the first day. That's why digital evolution provides those who are incorporating it with freedom and peace to operate in a low- or even minimal-risk area. When implemented the right way, these transformations are risk-free (Figure 2.4).

Flexible

Digital evolution allows companies to set short-term goals instead of being distracted only by the end goal of their transformation. More importantly, companies are constantly in control of their budget and time, allowing them to be flexible as much as they want to at any given moment.

Successful evolution

As evolution always continues and companies are constantly improving their processes to reach better business results, it might be challenging to

Figure 2.4 Universal API illustration.

determine whether the evolution is being successful or not. However, in that lies its true value for all companies. If the transformation is not providing a company with the desired results, its business leaders can determine what needs to change to steer the company in the right way.

There will always be something worth transforming digitally in the organization, and until all efforts are focused on making the business work better, digital transformation will be successful. To ensure that the organization is on the right path, its business leaders need to keep in mind two things: determining the approach and availability of skills.

Determining the approach

Determining the right goal of a company's digital transformation is vital; however, it will be impossible to achieve any business goals without determining the approach for it. The best way to ensure that the end goal is ultimately achieved is by breaking it into smaller, more achievable goals. When focused more on the set of goals that bring the company closer to its end objective, the idea of revolution becomes replaced with evolution. With each goal being able to reach with the help of technology, companies that recognize the value of IT will simplify their processes and make them more valuable for the entire organization, not just the technology department.

Availability of skills

Digital transformation requires certain skills a company's teams must possess to move in the direction of digital transformation. Without skills being available whether it's in-house or through hiring a remote team, companies will not be able to implement their determined approach and ultimately, achieve the goal they've set for their business. There are seven high-demand technology skills needed to implement the digital transformation process in any organization:

- **Cloud computing.** The cloud model enables organizations to scale their businesses and optimize infrastructure costs.
- **Mobile app development.** As the value of the global mobile application market is growing, companies have another powerful way to engage with their audience.
- **UX design.** With a growing number of mobile devices, there is a growing need for valuable digital content that has to be aligned with their customer's needs.
- **Blockchain.** Numerous industries are implementing blockchain-based technologies to improve their processes.
- **Cybersecurity.** With companies moving to the cloud, cybersecurity becomes a crucial aspect of protecting the company and everyone it interacts with, from employees to leads.

- **DevOps.** The entire digital transformation process requires a smooth software development process.
- **Artificial intelligence and machine learning.** The implementation of these two technologies provides vast possibilities to optimize processes and improve efficiency.

Evolving systems of digital transformation

The reason behind the need to make digital transformation a priority for all companies across every industry is the urgency of marketplace competition. Adding to that cloud-based digital transformation that is introducing new products almost every few days, there is no doubt that we're in the middle of the evolution of digital transformation.

Indeed, many companies have already created a new C-suite title, the chief digital transformation officer or CDTO. The purpose of this role is to determine which of all these new solutions and projects should be implemented in the company's everyday business and how. With or without CDTO, it can be a bit overwhelming choosing the system for the company as many requirements need to be met. Companies need to ensure they are taking a modern approach but also meeting the needs of their employees and customers.

Due to the evolution in digital transformation in the business world, companies first need to take a moment to review all solutions available to them at the given moment, select the right system for them, and then determine how to implement the system. Having a clear understanding of all solution types is the first step towards a successful digital transformation and all benefits that follow.

Systems of record

The way a company will start its digital transformation journey will depend entirely on the needs of that business. For companies that have started their transformation early, systems of record were the foundation of their digital transformation. Systems of record or SOR rely on utilizing internal data to store information that will be easily accessed by employees who need it.

A SOR example can include a CRM for a customer-facing business or an electronic health record system for a large health care institution. Systems of record are called like this because they create a record and specialized data a company can then use to make more informed decisions for the business. This is also the first wave of digital transformation because companies realized the potential of data and started seeking an efficient way to make the most of that data and enhance their relevant business processes.

Today, systems of record are still a central piece of the digitization process of any organization. Every business will need to rely on a SOR at some point to become more efficient; however, having a SOR will not be enough.

Organizations cannot be modernized just by having a solution; it needs to be implemented the right way to make the change an organization was looking for.

Systems of collaboration

After systems of record, what followed was systems of collaboration. The SOR didn't live up to its expectations as the data was only available for certain departments of an organization, and there was no simple way to disperse it and ensure everyone is informed. As companies were kept siloed, systems of collaboration offered a solution of tapping into that data gathered by the SOR and enabling for a flow of knowledge between departments at a speed impossible to imagine before this technology.

The best examples of platforms that are based on systems of collaboration are Slack and Microsoft Teams which allow teams across the world to communicate and collaborate on different projects. These systems of collaboration also allowed to pull out the knowledge that would stay hidden without them.

Systems of engagement

The next step in the evolution of digital transformation was systems of engagement. The previous systems, systems of records, and systems of collaboration gathered an immense amount of information that needs to be compiled and digested in the right way. And to do that, companies were in a need of yet another system. With systems of engagement, advancement in database optimization technology was achieved, so companies were able to gather and source data faster and provide valuable business insights at an impressive pace.

There are many platforms we use today that are based on systems of engagement, such as Zendesk, Constant Contact, and Facebook. The way that collected data is being used with such platforms has never been possible before the systems of engagement. These systems provided leaders and organizations with an ability to quickly process and make business decisions even quicker, turning them to be a crucial role in developing customer service innovation.

Simply put, businesses leaders were now able to develop digital strategies and replace the product-focused approach with a service-focused one. This resulted in an overall enhancement of the customer experience and prepared the business world for the following steps in the evolution.

Systems of productivity and outcomes

The last stage of the digital transformation evolution, or the currently best way to transform businesses, involves threading productivity and outcome systems. When organizations begin combining digitization with productivity and outcomes, all of their existing systems can be used to work together aiming to add more efficiency to their teams, but without adding more to the current workload.

Finally, we're seeing a progressing technology where businesses can make the most of all the systems available and use them in a unified way to create an improved customer experience platform that delivers revenue and productivity outcomes for everyone – customers, employees, and the company in general. Many companies now provide teams with the ability to link workflows and scale their productivity across the entire organization.

When a company achieves holistic digitization, all the workflows in the company can be mapped out easily, new tools can be adopted quicker, and the company can focus on creating a valuable end-user experience that leads to lucrative business results. The system an organization is using needs to be aligned with the quarterly goals and business outcomes to make it work. These systems are the most efficient way to make use of everything such technology has to offer, and there is an indefinite number of possibilities with it.

THE VALUE OF INFORMATION IN DIGITAL TRANSFORMATION

To understand the value of information in digital transformation, it is necessary to go through the three stages of the Web. Web 1.0 was focused on connecting information, whereas Web 2.0 also added an interactive element by using a range of technologies. Finally, Web 3.0 introduced us to the Internet of Things, where devices and artificial intelligence are connected.

To connect the systems of digital transformation, companies will need to connect information first. This goes beyond adding a few features as it's a rather challenging task in a complicated ecosystem of myriad formats, numerous standards, enhancing volumes, and the acquisition of devices to the entire digital space.

Digital transformation intertwines with information activation. After all, the digital transformation process in the context of digitization refers to the transformation of the physical to digital information. The objective of information activation is to deliver smarter and more valuable outcomes for businesses.

This requires a holistic and hyper-connected optimization for any business considering implementing digital transformation as we know it today. Holistic optimization is the direction in which all transformations will go; however, it will not be as focused on technology as it will be on the human and emotional aspects.

DIGITAL TRANSFORMATION VS. DIGITAL EVOLUTION

Is digital evolution just a synonym for digital transformation or do these two terms imply something completely different? When looking into the development of digital transformation, it becomes evident that we're talking about evolution. To continuously increment improvement is the definition of evolution. Yet, with the set of today's technologies and tools, companies can make transformative changes to their processes.

Digital transformation refers to using intelligence to drive transformative business outcomes. However, the transformation that is a result of all these systems doesn't necessarily have to be digital. No doubt implementing technology into a company's business operations can provide many benefits, but does digital transformation stop there? What happens when an organization uses new technologies to augment human workers with greater information, more profound insight, or a more thorough process to amplify their existing intelligence? It seems that digital transformation has already extended beyond the borders of the digital form, and maybe the next step of digital transformation will be changing the "digital" into "intelligent".

After all, as much as digital transformation is based on technology, there needs to be awareness of how worthless any new technology will be if it's not being used in a way that it delivers constant business value to organizations. The capability of technology is what delivers value, not technology itself. Even if the next step of this evolution is an enhancement of all technologies we know today, it doesn't necessarily mean that all organizations will enhance their businesses.

Looking it that way, it's irrelevant whether we call it digital transformation digital evolution, or even intelligent transformation. The name becomes irrelevant in front of the meaning it has for the business world. This evolution involves transforming the digital information an organization has by using all the new technologies, tools, and techniques available to it.

There is another digital transformation term that is rather confusing to a wider audience, and that's "disruptive". In other areas, disruptive indicates a disturbance or a problem that has interrupted something, so why is the term "disruptive technology" considered positive? Many assume it's because the new technology will often disrupt the existing one; however, in digital transformation, the disruptive is not the aspect to be celebrated, incremental efficiency is.

The value of technology is not in disrupting the previous technology but in how much new value it can provide to those who will use it. Digital transformation also welcomes using existing and new technologies if the ultimate goal will be achieved easily. Similar to many other terms coined in the IT world, their meaning often evolves and outgrows the name.

Chapter 3

Digital transformation benefits

When integrating digital technology into all relevant business areas, fundamental changes will occur in how an organization operates its business. Many industries are already benefiting from digital transformation, from modernizing legacy processes and accelerating efficient workflow to increasing profitability and strengthening security.

Most companies today are working in the cloud, but the true value of digital transformation goes beyond data migration. It can spread across the entire organization and create a technology framework converting services and data into actionable insights that enhance every organizational facet.

Instead of just migrating organization data to the cloud, businesses should begin reevaluating and optimizing their systems and processes and making them interoperable and flexible to support business intelligence and provide the organization with everything it needs to succeed. A digital transformation process will affect every level of an organization and gather all the valuable data allowing teams to work together more efficiently.

When organizations add workflow automation and advanced processing, they can strengthen the customer journey and provide more value. Organizations can now generate immense value from transforming the way they work.

RECENT EVENTS

The COVID-19 pandemic has irretrievably changed the world as we know it, and all organizations had to adjust as well. Those businesses that began their digital transformation before the pandemic were able to make rapid adjustments by relying on the cloud architecture, latest security protocols, and the entire range of robust technologies to support the remote-work environments and virtualized business-related transactions and interactions.

On the other side, non-profit and governmental organizations struggle to adjust to the rapidly changing economic and social conditions. Those that managed to provide rapid answers to the new environment were the ones

DOI: 10.1201/9781003305163-3

that succeeded and those that haven't noticed negative trends in their business numbers.

The market is now changing more rapidly. More changes are happening quicker than 10 or 15 years ago. Businesses that are more mature in their digital transformation process have witnessed much less disruption and saved money because of it. In the wake of the so-called COVID-19 disruption, business leaders have decided to accelerate their digital transformation initiatives and are aware that more changes are coming their way.

Organizations have the opportunity to work faster and more innovatively. Due to cloud computing and mobile technologies, they can provide reliable access to crucial business platforms from anywhere. Not to mention that such automation will also speed up business processes, decrease errors and focus employees' attention on higher-value actions instead of basic repeatable tasks.

When adding artificial intelligence, the value for businesses can augment tremendously. Employees will work more efficiently and organizations will improve how they interact with their audiences, leading to more sales and profitability.

As lowering operational costs is one of the main drivers for digital transformation, it is essential to be aware of its capability to reduce hardware and software costs, among other digital transformation benefits. As organizations continue transforming the way they do business, new technologies will appear, leading to undiscovered benefits for organizations.

DATA COLLECTION

Each business relies on its customer data to excel. Gathering them was an issue that was easily solved with Google Drive, Dropbox, or any other cloud-supported platform. The value organizations have from data is not in gathering, but optimizing them for analysis that drives their business forward. With digital transformation, organizations have a data-gathering system that also incorporates it entirely for business intelligence at a much higher level.

Digital transformation also helps these organizations to translate raw data into valuable insights that show an overview of the customer journey, operations, finance, and production. To constantly transform how customer data is collected, stored, and shared, it is needed to evaluate data collection and optimization processes frequently.

Without a doubt, data management still prevails to be the core challenge for many businesses considering implementing digital transformation. Understanding how to enable, activate, and manage the right data is more difficult than it seems at first. Evaluating the customer journey as part of the digital transformation process helps organizations give their clients higher autonomy over their data and demonstrate how crucial data privacy is for them. With customers becoming increasingly aware of how their data is being used, businesses must collect and manage data in a way that favors their customers. That is why many organizations have already developed their data strategy.

Having a defined data strategy in place is vital to solving fragmentation that typically occurs with organizations gathering a large volume of data. The main elements of a successful data strategy involve data, technology, analytics, and delivery services that lead to higher reach, revenue, and return. For organizations implementing digital transformation, a data strategy serves as a roadmap to link all data, both online and offline, to define steps required for enhanced omnichannel marketing and customer engagement.

To solve data fragmentation, organizations need a unified data layer, which is where most digital transformation falls apart. That is why it is necessary to build a quality data strategy within a digital transformation plan. A unified data layer is an open, trusted framework for data that provides companies with an omnichannel customer view that is essential when seeking opportunities for sustainable growth.

RESOURCE MANAGEMENT

In digital transformation, information and resources are converged into a suite of business tools. Instead of having a range of software solutions and databases, resource management consolidates all company resources and decreases vendor overlap. On average, an enterprise business will use several hundreds of applications. These applications are not integrated into one process that demonstrates a clear overview of all business operations, and this is exactly what digital transformation can do.

Digital transformation aims to integrate all business applications, databases, and software pieces into a centralized repository. This funnels all organizational workflows into one and saves a tremendous amount of time employees spend migrating data and managing processes across different workspaces. It also reduces errors as everything is stored in one place.

Besides technology, other priorities in digital transformation are efficiency, communication, and customer-centric approach. All three of them can be delivered with resource management. Just like digital transformation, the objective of resource management is to help any organization adapt to a constantly evolving business landscape. This is done by optimizing how managers prioritize projects and tasks, manage resources, and lastly, implement certain technologies.

Besides a defined data strategy, quality resource management practices should be an integral part of any digital transformation strategy. These practices allow organizations to prioritize all initiatives around digital transformation efficiently, meaning that the crucial areas are implemented first. This prepares organizations for a smooth transition that results in optimal returns on investments by strengthening workplace productivity, resource management, and a customer-oriented approach.

For a resource management strategy to succeed, it will need to have these three key components:

- **Collaboration.** Every organization's members should work with others to ensure that the shared vision is achieved.
- **Responsibility.** The organization's members should be involved in the planning, implementation, and decision-making processes.
- **Flexibility.** As business and technologies are constantly evolving, organizations need a dynamic strategy that yields growth and development.

A digital transformation initiative can only excel if the organization's teams and resources are well managed. As much as the large organizations are considered to be actors of poor project management and performance, there is a lack of quality in this area in almost every business, regardless of its size. That is why resource management should be a top priority for organizations implementing digital transformation or any other vital business initiative.

Four actionable solutions of resource management

Organizations looking to improve digital transformation through resource management should focus on four actionable solutions leading to desirable results.

Measuring the ongoing digital transformation performance

Access to accurate data and real-time analytics is essential when planning and implementing any digital transformation strategy organization has decided. There is immense value in measuring the impact a digital transformation initiative has had on its organization and determining if it helped achieve the business outcomes it was supposed to. When measuring the ongoing performance of organizational efforts to implement a digital transformation strategy, business leaders can easily optimize it and maintain long-term growth and development.

Implementing effective resource management solutions helps businesses manage and use these resources more efficiently, which is crucial for digital transformation success. To ensure quality resource management, initiatives have been implemented, businesses leaders must cover the following:

- Provide their employees with a platform that is simple to use for managing tasks, projects, team members, and assets to simplify and streamline the entire digital transformation process,
- Help all project managers involved make better-informed decisions around allocating work and resources,

- Enable managers to map out action plans and break these actions into prioritized tasks needed to achieve digital transformation.

Transparency around digital transformation

When implementing a digital transformation strategy, organizations should have one crucial goal – creating holistic workplace transparency throughout all levels and hierarchies. People within the workspace will be the ones that have the needed experience and technical knowledge, which makes their opinion valuable when planning and optimizing the digital transformation strategy. Also, organizations must ensure that all executive teams are included in the strategy as soon as the initial planning stage has terminated.

By keeping everyone in the loop during the entire transformation journey, the level of effectiveness will constantly grow. Allowing employees to have an open line of communication for any feedback or reporting issues, securing that leadership is updated and well aware of performance at the same time, helps everyone involved make better-informed business decisions. The outcome is an organization able to collaborate cohesively to achieve goals on all levels and hierarchies.

Optimizing collaboration and productivity through technology

A fast-paced environment in which businesses operate today makes resource management solutions necessary to guarantee optimal collaboration, especially for larger organizations managing multiple projects. Both digital work and resource management will enhance any transformation initiative by:

- Creating a clear communication channel for all stakeholders, managers, and employees can help discuss all project information in real time without being restricted with internal silos,
- Supplying a visual representation of project workflow from beginning to end,
- Simplifying the process of identification and correction of any arisen problems for managers.

In the digital era, being adaptable distinguishes successful organizations from unsuccessful ones. Those who can quickly implement new strategies when needed and collaborate across their internal and remote teams will experience digital transformation success. As tech resistance is one of the most common digital transformation errors organizations make, it is evident how vital is quality resource management. Businesses embracing software solutions that improve team collaboration and communication will more likely adopt a unified approach towards simplifying and streamlining digital transformation. This makes it much easier for these organizations to achieve their business objectives.

Becoming agile

Undoubtedly, agility is what drives digital transformation. Businesses that are agile and dynamic in their operations are a better fit to facilitate any digital transformation initiative effectively. The reason for that lies in agility tearing down bureaucracy. This can be achieved by the following:

- Stepping away from the traditional hierarchical structures and choosing a team-oriented approach
- Eliminating the focus only on conventional metrics indicating profitability and shifting to finding efficient ways an organization can cope with dynamic customer preferences
- Embracing a broader range of strategic execution methodologies

Organizations that can quickly shift their strategies, change objectives, and adopt new technologies will easily align their transformation strategy with the market expectations. An approach that is based on agility will replace a milestone-driven mindset with continuous delivery. Such shift will result in managers, stakeholders, and employees being better equipped to satisfy any growing digital transformation needs in a dynamic and very competitive environment.

Blending it all

These four actionable solutions of resource management shouldn't be singled out but blended in. The way an organization manages work and resources plays a key role in determining the triumph of any digital transformation initiative. The ultimate objective all businesses should strive to achieve is complete transparency.

With full transparency, it becomes easier to assess both strengths and weaknesses of digital transformation strategy accurately because the immediate priorities, issues preventing the completion, and ways to improve cross-organizational collaboration become clearer. When doing digital transformation like this, organizations gain deeper insight into how it can be utilized to improve their overall performance.

DATA-DRIVEN CUSTOMER INSIGHTS

To fully understand and take action from customer insights, they need to be data-driven. Once organizations begin getting a clear understanding of their customer needs, they can develop a business strategy that is based on a customer-centric approach. Any type of data can be valuable for an organization, from unstructured data like social media metrics to personal customer information. The more data a company has, the more possibility it has to drive sustainable business growth and optimize all areas (Figure 3.1).

Figure 3.1 Data driven society.

Moreover, quality data also leads to more relevant, interactive, and personalized content for customers. After all, today's customers need convincing from businesses to buy from them. If an organization doesn't know how to sell its products or services through words, its sales numbers will suffer. Without a doubt, organizations that know how to leverage data to drive digital transformation will prosper higher than their competitors who don't.

Business leaders need to possess a detailed understanding of their audience. This includes dynamic demographics, recent behavior patterns, interactions, and expectations, which are then incorporated in the design of new company products, services, and value propositions. Digitalization has surpassed just conventional marketing technologies and mobile and social platforms and expanded to multimedia, IoT, cloud, automation, and artificial intelligence. All these technologies are utilized to redefine customer value propositions.

When an organization redefines customer value into products and services, digital technologies can be used to deliver incredibly valuable experiences to customers, leading to higher customer satisfaction and operational efficiency. Besides revolutionizing business models, products, and processes, digital transformation irretrievably changes leadership style, organizational structure, and company culture. Those who manage to master these fundamental elements will guide their organizations to become more valuable, competitive, and profitable in the digital age.

Transforming companies with data

The transformation of a company will always greatly depend on the behavior of its leader, which consists of assumptions, individual perspectives, personal experiences, and risk profiles. Personal human factors can impact the way a company embraces digital transformation and differences between leaders can be devastating for company success. Herein comes the immense potential of data.

Data beats assumptions, hunches, and personal bias. Data is not affected by personal human factors, and it can only be read in one way. Leveraging data brings objectivity and an intrinsic commitment to priorities across the entire organization. It also helps business leaders manifest a comprehensive, customer-driven vision and strategy that will produce the desired results because they are based on technology that studies customer behavior in detail.

Data leads to competitive advantage

Digital maturity has gotten to the point that many organizations are generating business growth from their digital transformation processes. Implementing data-driven customer insights into business strategies ultimately generates more success because today's customers know what they want and they buy from those who give them just that.

Translating customer insights into business outcomes will separate successful organizations from unsuccessful ones. Essentially, digital transformation is becoming the foundation of new business models, combining unified technology systems with the engine of growth and data as its fuel. As data-driven business models are being implemented across all industries, it's crucial to pay more attention to data management.

Data is only as valuable as its management across the entire organization. There is still a lack of data and analytics outside the IT function. Data management and blending are factors that most companies miss when implementing digital transformation. In other words, being aware of data value is not the same as leveraging that data in a way that generates business growth.

ENHANCING CUSTOMER EXPERIENCE

With digital transformation, organizations are unlocking efficiencies in their teams, but they are also delivering more intuitive experiences for their customers. It goes from email communication to digital products and user portals, and even the way you reach out to your new prospects. Today's customers have elevated expectations for digital experiences. They are used to having numerous choices, reduced prices, and quick delivery. Customer experience is the new hot area for brands in most industries, and organizations are becoming increasingly aware of it.

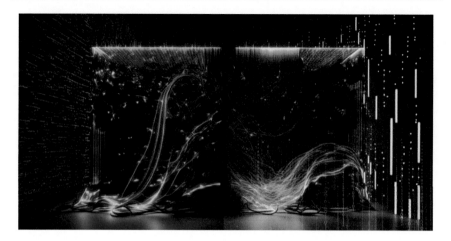

Figure 3.2 **Customer experience.**

Customer experience is becoming the key driver of sustainable growth for most businesses. By enhancing how customers interact with their businesses, larger organizations can increase their annual growth by millions of dollars (Figure 3.2).

One of the most concerning aspects of customer experience is customer data privacy. Those who successfully demonstrate how they value their customers' privacy can differentiate themselves from the overwhelming competition. Customers must be in control of the way their data is being collected and used, but also empowered to make any further decisions around their given data.

Digitally conscious customers

Digital transformation is based on digital technologies, which have completely transformed customer habits. The evolvement of mobile devices, endless number of apps, automation, and machine learning enable customers to get precisely what they want at the moment they want it. Moreover, these digital technologies have triggered a shift in customer expectations, creating a new type of modern buyer. Customers of today are continuously connected and aware of the potential that lies behind these technologies.

Due to numerous opportunities that appear from using these new technologies, customers often evaluate organizations on the digital customer experience first. The digital-first approach challenges organizations to rethink the existing ways of interacting with their customers and adding more value to them.

From a sales perspective, this implies replacing cold calling with social selling. Customers are active on social media, and that is where organizations need to be as well to sell. Rather than waiting for a customer or prospect to make the first contact, organizations should reach out to them first. This change allows organizations to build deeper relationships and educate their customers about their business and the products or services they are selling.

From a marketing perspective, the digital-first approach requires reducing the amount organizations are spending on offline marketing activities – direct mail, billboards, and TV ads. Customers are expecting highly targeted messages, which can only be accomplished through a data-driven marketing strategy. Organizations need to use all these digital channels to implement search engine marketing, email marketing, and account-based marketing strategies.

From a customer service perspective, teams no longer have to wait for the phone to ring or fax to come through. With the digital-first approach, companies are becoming proactive instead of just reactive. Customer service teams need to be proactive in ways they help their customers, keeping in mind that they use a range of digital channels to look for support. Now, social media, forums, review sites, and online communities are part of the customer service ecosystem.

Reimaging the entire customer journey

Customers are becoming increasingly in control over how companies deliver experiences, meaning that every new experience a company builds should meet the customer demands. As every experience with an organization impacts a customer's overall perception of the business, all these organizations must apply a customer-centric approach throughout all their business strategies. More importantly, this shift to digital allows businesses to create enhanced relationships with their customers.

To do so, they must create an elegant, flexible IT environment. Having the needed technology to make the most of digital strategies is essential for success in today's business world. That is why organizations across many industries are recognizing the need for agile systems, in which cloud technologies stand out as the most implemented solution. Cloud technology allows organizations to be faster, more dynamic, and flexible, providing them with the opportunity to test new cost-effective and low-risk projects. When implemented the right way, this technology helps tremendously to meet customer demands faster.

However, even the best data-driven business strategy will not produce the desired results if it's not based on personalization, one of the vital factors in digital transformation. Customers want organizations to treat them as unique individuals and know all about their personal preferences and recent purchase history. There are three aspects of personalization that should be a part of any digital transformation strategy:

- Recognizing all customers by their name,
- Having all information about customer's purchase history,

- Recommending products suitable for them and based on their recent purchases.

Personalization is the value customers expect from organizations that collect their data. Knowing that providing their data will allow companies to provide a better customer experience creates a positive change in customers' perspectives.

ENCOURAGING DIGITAL CULTURE

When employees are given the right tools, which are tailored to the organization's environment, digital transformation can encourage a digital culture. Inevitably, tools that allow employees a seamless team collaboration and increased efficiency are also responsible for moving the whole organization ahead, especially digitally. Digital culture is critical for organizations to remain sustainable. It pushes the upskilling and digital learning among employees to maximize all digital transformation benefits.

Digital culture

Digital culture can best be described as a concept that defines the way technology and the internet are shaping the way we interact. It impacts the way we behave, think, and communicate with other humans and as a society in general. Digital culture is the result of persuasive technology and disruptive technological innovations that follow it. It can be applied to numerous topics, but it always revolves around one main point – the relationship between humans and technologies.

There are a few reasons why organizations implementing digital transformation can benefit from digital culture:

- **Breaking hierarchy:** Incredible value lies in allowing employees to make their judgments which are not limited to hierarchy.
- **Accelerating work:** With better judgments, employees can make quicker and better decisions.
- **Stimulates innovation:** Organizations become workplaces where employees are motivated to try and learn new things.
- **Attracting new generations:** Millennials and GenZ seek environments that don't offer traditional 9 to 5 jobs but are autonomous and collaborative instead.

More than talking about changes in actions, digital culture focuses on changes in the way these organizations think about business. Changing the mindset is essential when talking about digital culture in businesses implementing digital transformation. For instance, understanding that a report cannot provide as valuable customer feedback as the customers themselves is one of such changes. A change in mindset leads to changes in actions, so these businesses will start making decisions in real time based on the data gathered from the customers.

Without a doubt, building a digital culture is challenging, but it's the only way to quality technology adoption. Digital culture also takes time and requires to be accepted company-wide and employee-wide. That is why organizations must have a clear vision and strategy for the way these digital changes will be implemented, including defined goals, to ensure everyone is moving in the right direction.

The organization's employees are at the heart of digital culture, so there must be constant two-way communication. Business leaders should keep employees informed about the digital culture at all times but also ask for their opinion in the building and implementing phase. On the other hand, employees must actively participate in creating the digital culture in their organizations by learning new things and engaging more with their colleagues and executives. If employees are confused about their role in the transformation process, the digital culture cannot be built.

True digital changes will occur when all levels of the organization are aligned with the digital transformation strategy and can participate in it in their way. If everything else stays the same and organizations push for digital culture implementation only during meetings and presentations, employees will feel the disconnect between the organization's words and actions and the change will quickly hit the wall.

All decisions must lead to changes in the day-to-day operations, but they should also be made differently than they used to. By focusing on digital culture, organizations will naturally start changing the ways they do business. An organization must support innovation at all levels without allowing leaders to squash change and agility. That said, digital transformation is a coin with two sides and it doesn't have any value without both of them – technology and culture.

IMPROVING EFFICIENCY AND PROFITABILITY

Companies that begin with digital transformation report improvement in their efficiency and profitability. Committing to digital pays off in the long run, so companies that have already started with their digital transformation process report an increase in their profitability numbers. As talent management and development are considered key drivers of digital growth, one of the outcomes will inevitably be improved efficiency and profitability.

Companies are increasingly focusing on talent attraction and retention, knowing that employees are crucial for sustainable business growth. Another vital element is an investment in the organization's digital skills, which is expected to steadily increase revenue in the next few years. This conclusion comes from comparing organizations that have implemented digital strategies and organizations that have yet to adopt them.

Their main difference is in the talent management approach. The leaders of digitally transformed organizations anticipate that digitization will change talent management, as opposed to their counterparts who are not putting any focus on it. More importantly, digitally transformed organizations understand that digital transformation has simplified the process of talent attractions and retainment.

Businesses that will lead tomorrow are the ones putting their employees at the center of their digital strategies. A successful digitalization depends on employees who work in innovative and forward-looking organizations devoted to investing in their people to ensure they have all needed to face the challenges of tomorrow.

Employee engagement can also be improved when the digital transformation is completed. Once the digital strategy is fully implemented, the employees in the organizations tend to become more engaged. This is because they are being heard during the moment of changes, and their opinions matter to the organization. If these changes occur without a proper dialogue with the organization's employees, the level of engagement will be either the same or even lower than before the transformation.

Organizations implementing digital strategies must also be aware of the potential need for new roles that haven't existed before. Businesses should prepare their resources to create new roles that are aligned with their technological goals. These roles will be focused on maximizing the potential of any digital transformation project, while also growing the revenue. After all, there is no value in digital transformation without the people using it.

The biggest value of digital transformation lies in the employees. The way they work, things they know, and the skills they possess to face their dynamic workplace are what determine whether a transformation has been successful or not. As most companies have only started to realize the immense value of human factors, investing in digital skills is what will differentiate successful organizations from the ones that will stay behind.

Key people across an organization are what is driving a business to success, but more importantly, it gives consistent worth to these organizations which need human factors to connect with their customers and reach their business targets. Understanding, implementing, supporting, and making the most of any type of change begins and ends with employees. Organizations that fail to align their digital transformation processes with their employees will not be able to keep up with their competitors and ultimately, lose the race.

MORE AGILITY

Organizations are increasing their agility with digital transformation as they enhance speed-to-market and implement Continuous Improvement strategies, policies that help organizations focus on improving regular tasks and projects through regular incremental enhancements or larger process enhancements. With more agility, organizations have more opportunities for faster innovation and adaption, but also generally improving the business.

Agile organizations will gain a competitive advantage more easily than the ones that are not. As digital transformation implies a lot of changes on many levels, agility ensures that businesses have what it takes to adapt and continue moving towards their objective. What most businesses lack is implementing agility on a holistic organization level instead of sectional level of it. Having agile technology or project management team is important; however, it will not suffice in today's competitive business world. By being more agile on all levels, organizations can minimize risks and ensure planned outcomes are always achieved.

The reason why most organizations haven't implemented agility into their business methodology is the fear of failure. Business leaders are taking a set of actions to avoid business failure, so embracing it is not a part of the company culture as we know it. Yet, to become an agile organization that completed the digital transformation successfully, it is required to accept failure as a valuable part of the entire process.

In digital transformation, failure becomes 'testing and learning'. By incorporating agility, organizations accept the unknown and are ready to discover whether the new ways and ideas work or not. This approach is hugely beneficial for organizations that seek innovative ways of doing things and are not afraid of failure. The fear of failure results in business leaders bending the truth to present apparent success when the project's objective hasn't been achieved. Besides the lack of transparency, this makes organizations continue with projects that don't work, without even realizing it.

As businesses are becoming more open to the 'test and learn' approach, there is a new trend of transparency which leads to all stakeholders feeling more involved and responsible for the success of a project. Agility puts an end to embellished performance reports that are causing harm to organizations slowly but steadily and creates a healthier environment where everyone involved is informed about the performance in real time and can take improvement actions at any given moment.

Resistance to agility is often justified with high costs and the possibility that the 'new' cannot replace the 'current', but the costs of risks that appear due to lack of agility can be even higher and more harmful for the organization as they cannot be predicted. Agility is necessary for businesses to rethink their existing strategies and adapt to new challenges in such a dynamic environment. Without it, they will experience lesser success and the objectives from their strategies will ultimately not be achieved.

All organizations need to take their first step towards agility by understanding that failure is a part of the learning experience and shouldn't be considered as a negative consequence. They can do so by revising the terminology that is placed deeply in their company culture. For instance, terms like 'failure' or 'unsuccessful' should be replaced with 'piloting', 'learnings', and 'exploration'.

Instead of having two potential outcomes of each project – success and failure, organizations should start focusing more on 'test and learn' phases throughout their entire project planning. Such testing involves both technical testing and employee and customer feedback and consultations, risk analysis, and more. Expanding the understanding of 'testing' to more than just technical projects, and seeing its value when implemented on the corporate level, is where organizations should seek drivers for digital transformation success.

Exploring and reacting to learnings gained from testing is another aspect lacking in project planning. Agility requires time, and skipping any of these steps will not provide organizations with valuable learnings that lead to success. These learning are crucial for future projects and will help employees enhance their knowledge and skills needed for a thriving business.

Finally, by taking an agile approach to digital transformation, organizations focus on setting a goal, but not the time or costs required. These projects are broken down into sprints, where each sprint has an individual goal, and there is enough time to make all needed adjustments to ensure that the desired outcome is achieved. Because these sprints are aligned with the core project objective, they can easier be completed. They contribute to the overall success of the project even more than just taking actions in a wider strategy that lacks comprehensive planning and outlining of milestones needed to bring the business closer to the objective.

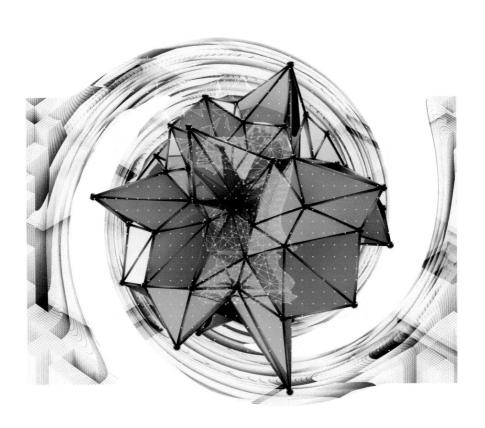

Chapter 4

Digital transformation strategies

A digital transformation strategy is a high-level tool that outlines an organization's objectives and digital initiatives. It refers to the means of the use of technology to complete tasks that were done manually before. At a higher organizational level, digital transformation is an opportunity to completely reshape the business by aligning the existing operating model with new ones that are appearing.

WHAT IS A DIGITAL TRANSFORMATION STRATEGY?

In a highly competitive environment, businesses cannot succeed without a strategy. When it comes to digital transformation, digitizing everything is simply not enough. As a fairly new concept, digital transformation is something most organizations still don't understand profoundly, making it even more challenging to create a valuable strategy. That said, not every plan can be called a strategy. To-do lists, resource plans, operational plans, and capital investments are not strategies (Figure 4.1).

A valuable strategy must consist of the following components:

- Diagnosis,
- Guiding policy,
- Coherent plan of actions.

Considering these components, a digital transformation strategy refers to a plan of action to reposition an organization in the digital economy. Such a strategy will be based on data to detect new opportunities and conduct a plan of actions to achieve them. The long-term goal of a digital transformation strategy is to continuously leverage emerging technologies and maximize new opportunities with proper skills and mindset.

When working on their strategy, organizations must be aware that digital transformation will not be top-down or bottom-up, but side-to-side. These changes are complex, so they cannot happen all at once or be willed into

DOI: 10.1201/9781003305163-4

Figure 4.1 DT strategy.

existence. For digital initiatives to be implemented successfully, employees will need to understand and embrace them.

As data is key for a digital transformation strategy, businesses must change how they seek opportunities. This means their decisions should only be data-driven to excel. Here is where the value of strategy comes in, exhausting the data to find numerous opportunities for business growth. When based on data, a digital transformation strategy guides organizations through change and gets them to the digital side.

An organization can put as many resources into the digital transformation implementation, but there will be no digital transformation without an in-depth strategy. Organizations can have the best technology but lack the right people; or, have enough resources but lack a good understanding of data. Each of these components is crucial for any digital transformation strategy, and without them, a business will only be able to digitize instead of fully transform.

COMPONENTS OF A DIGITAL TRANSFORMATION STRATEGY

Four components will need to be covered in a digital transformation strategy and incorporated into the business mindset as well. If people don't believe in the changes they are implementing, the digital transformation strategy will fail. That is why all organizations implementing a digital transformation strategy should equally focus on developing a quality strategy and changing the business mindset.

Communications

Changes a strategy aims to implement should be communicated to everyone inside and outside the organization. The strategy must respond to 'what will be said', 'how it will be said', and 'what is the value it will create'. All audiences, including employees and customers, should be educated about the value these changes bring to the business. Even the most efficient and profitable changes will fail in the long run because their value lacks proper communication towards the employees, customers, or other stakeholders.

That said, these changes should be big enough to be considered transformational. Minimal changes, such as a promotional video or a last-minute initiative, will not convince anyone. Organizations should communicate relevant changes that benefit the business as well as their employees and customers and not try to present small changes that will have little to no impact as something revolutionary.

Culture of innovation

Every digital transformation strategy has to address innovation and create the actions around it. After all, its ultimate objective is creating a culture of innovation. To constantly innovate something new, a high level of creativity is needed. This requires organizations to nurture creativity which is often marginalized. Innovative businesses are the ones that understand and focus on developing creativity, knowing that practicing creativity leads to innovations.

However, numerous businesses will mistake good ideas for innovation. Purchasing technology and implementing it is not enough to transform business, but it only makes an aspect of it more efficient. Innovation begins with a customer issue or pain, which the strategy has previously identified and aims to solve these issues or pains with a set of carefully thought actions.

Technology

Technology inspires innovators and allows new business models. Leaders need to understand the business potential of emerging technology. Their understanding doesn't have to be profound, but they need to possess knowledge about how this technology can bring more value to the organization.

Data

To have an overview of business problems, an organization will need data. Having access to valuable data and knowing how to use it help business leaders identify business problems, the foundation for any digital transformation strategy. With its main components – diagnosis, guiding policy, and coherent plan of actions – organizations can find the 'unknown' through data.

In a way, business problems reveal data and data reveals business problems. Once the business problems are known, the process of innovation begins. Submerged in data, any innovation and technology implementation processes result in more predictability, which is essential when strategizing.

STEPS OF DIGITAL TRANSFORMATION STRATEGY

Besides incorporating these four components in their in-depth digital transformation strategy, organizations must follow steps to ensure business success. These steps help companies meet their objectives through a set of detailed actions and their audience embracing the digital initiatives it involves.

Business assessment

Each strategy should start from assessing the current business in detail. A comprehensive initial assessment should cover the strategic goals, key performance indicators (KPIs), and growth opportunities an organization has.

Researching the industry and competitors

Once the business is assessed, an organization should compare itself with its direct competitors and the industry. Have the competitors already started the process of digital transformation? Which digital solutions have they implemented? Which emerging technologies or tools are used within the industry? Conducting a competitive analysis helps organizations understand the market and identify opportunities to maintain a successful business.

Prioritizing digital initiatives

The digital initiative of one organization will be significantly different than the one their competitor has. This is because each business works on its initiative by determining its scope, resources, and return-on-investment (ROI). Prioritizing initiatives should be based on their value, estimated effort, and most importantly, their impact.

Creating a delivery plan

The organization's delivery plan should have a roadmap for all digital initiatives organization is planning to implement during digital transformation. The organization should also include resource requirements and criteria to measure results. Before they start incorporating the delivery plan, they should identify all development and delivery processes needed for forming standard practices that will apply to the entire organization.

Creating a budget plan

Most digital transformations fail because organizations don't have enough budget for their implementation. Creating a budget, anticipating delays, and planning for the emerging technology contributes to the overall strategy as those deciding on the actions will have enough resources to implement them well.

Creating a resource plan

However, technology is not the only digital transformation cost. Organizations need to assemble their teams with the right skills and mindset to put the strategy in motion. Besides that, they will need to assess their current abilities and the people they need to implement their digital initiatives.

PRINCIPLES OF DIGITAL TRANSFORMATION STRATEGY

Not only do the organizations need to cover four components and follow certain steps in their digital transformation strategy, but they also need to incorporate seven key principles if they wish to excel. These principles help organizations stay on track with their digital transformation efforts and achieve their objectives more easily.

Defining the reason for digital transformation

Within one organization, digital transformation will have different meanings for different people. The mistake that many business leaders make is they assume that digital transformation leads with technology and the rest of it will fall in place during the implementation. This approach often fails because it starts from the finish line. A successful digital transformation begins with identifying the needs and objectives and then building the strategy.

The strategic goals are a great start to any digital transformation strategy. What is the plan for the next five or ten years? This approach looks at digital transformation from a more efficient perspective than starting with technology. A clear business value needs to be identified from the beginning to implement any digital initiative.

Digital transformation is not a one-off project but a comprehensive, ongoing strategy that aims to position the organization for the future. The organization is set for success when building the digital transformation strategy on business priorities and unique business value.

PREPARING FOR CULTURAL CHANGE

Being enthusiastic about digital transformation is critical for cultural change that must occur over time. C-suite support, which includes CEOs, CIOs, and

CTOs, often determines the outcome of a digital transformation. Without their support, the organization will lack a person in charge of making decisions and ensuring the proper implementation.

Technology cannot make up for people missing to implement the digital initiative. As people are naturally resistant and skeptical towards changes, organizations need those leading others into digital transformation. They are *cheerleaders* of change due to their understanding of the digital transformation vision and benefits for everyone in the organization.

That is why as much as C-suite is crucial, it cannot stand alone amid the digital initiative. All departments and teams, including sales, design, and marketing, need to be prepared for cultural change by understanding what it implies and what it brings.

It is bit of utopia to expect no resistance among employees in digitalization process. Even managements in larger companies are willing to ask additional questions instead of trying to use new things. Start from yourselves. If someone would give you an Android instead iOS, how long would it take you to get used? Same and even worse with more resistance goes when talking about new processes. Mid-aged people could sometimes be "harder to crack" since they know how to do the job. They are used to one process. Only optimal approach in such cases is to open their eyes through education and what is the main goal of our digital transformation. By doing so we'll be having at least minimized resistance and hopefully after seeing the results it would be gone for good.

It is known that people's resistance can be hidden or postponed. Making digital transformation strategy familiar and clear is a way to make that resistance lower and involve our employees in each step of its implementation.

Starting small, but strategic

Digital transformation is a long-term journey that doesn't require completion to be considered successful. Digitally transformed companies can always find something to improve, optimize, or eliminate. Such transformation requires strategic thinking while focusing on more than completion.

On the other hand, its beginning is very clearly defined. Organizations must identify small and strategic actions putting digital transformation into action while motivating people to embrace the changes. These actions are often called "quick wins" as they lead organizations to success quickly and improve team morale. Quick wins usually show measurable results within six months.

That said, these quick wins must align with digital transformation. Mostly, these are actions derived from an overall strategy. Once the organization has defined its digital transformation strategy, it will break the strategy into actions that will help achieve the ultimate objective. The objective of a digital transformation is why it decided to start with digital transformation in the first place.

Mapping out technology implementation

When organizations add new technology to their existing operations, it doesn't result in digital transformation as much as in inexpensive, inefficient operations. When implementing one of the emerging technologies, it is essential to change the processes and cultures of the organization as well. Keeping in mind that technology is just one of the principles of digital transformation, organizations must have an in-depth strategy and informed, engaged people before jumping into technology implementation (Figure 4.2).

Once the organization has covered the first three principles, the technology becomes a needed tool to help it achieve its desired business objectives. While mapping out technology implementation, organizations must create a vision to build the business case for any digital initiative. These steps need to be simple and with measurable results.

Some of the most commonly used technologies across organizations when implementing technologies are as follows:

- **Mobile:** The number and usage of mobile devices is increasingly growing over the past years, and mobile technology became the cornerstone of the digital transformation strategy. Mobile enhances the speed and volume of interactions between an organization and its audience.
- **Internet of Things (IoT):** How organizations use data to enhance business makes the Internet of Things potent and the Internet of Transformation. The combination of data and intelligence strengthens innovation and transformations, resulting in the cloud, big data analysis, and other related technologies becoming key for digital initiatives.

Figure 4.2 Technology implementation.

- **Digital Twin:** This technology learns and updates itself constantly using real-time data to represent operating, working, and environmental conditions in almost real-time.
- **Cloud:** All types of emerging technologies, including private, public, and hybrid cloud platforms help create or modify business processes, culture, and all customer experiences to meet the market's needs.
- **Robotics:** As a perfect solution for replicating repetitive operations and performing time-sensitive tasks faster, robotics is a cost-effective solution that also reduces the number of human errors.
- **Artificial Intelligence & Machine Learning:** Artificial intelligence uses algorithms to create or modify programs to maximize the insights gained with machine learning. When using both of these technologies, organizations can gather and analyze data to better understand their customers.
- **Augmented Reality:** This enhanced version of the physical world presents an entirely new way of engaging with machines and performing tasks which is incredibly beneficial for organizations implementing digital transformation.

Using one of the mentioned technologies often will not be enough to achieve the wanted business outcomes. Simply using them will not suffice, as technologies an organization decided to utilize must align with the overall digital transformation strategy and have a clear purpose to fulfill. As these technology providers will often be located outside the organization, it is essential to choose technology partners wisely and ensure they are the right for organizational needs.

Seeking partners

When an organization has decided on the technology that will drive the digital transformation changes, it will need to start seeking its technology partner. That said, any products or services that will need to be outsourced should be considered in detail as they might affect the business outcome organization is striving to achieve. There are several questions that each organization seeking partners for their digital transformation must ask itself before committing.

One of the most important questions to answer is whether or not will the technology and the vendor support business scaling. If an organization and the vendor don't share a similar vision for a particular digital initiative, the results will be devastating. The partner should support the organization in its long-term strategy and have a clear idea about integrating these emerging technologies with the existing ones.

Organizations cannot implement even the best technology without the right partner. The chosen partner should have enough knowledge, experience, and background to be valuable to organizations seeking to digitally transform their businesses because they know what works and what doesn't. Such partners will also be efficient in solving problems across the entire organization, and not just one or two departments.

Gathering feedback for refinement

Organizations gather feedback not only to see what is the result of the implementation but also to improve it. Once technology partners have been chosen and the company has a clear digital transformation strategy to implement, it is essential to ensure that everyone is aware and accountable for what they need to deliver. Each of these deliveries is what will contribute to the digital transformation success.

At the same time, organizations will need to create a strong feedback platform where all stakeholders can learn from the experience as digital transformation is being implemented. Because digital transformation is just a journey, it consists of progress, adjustments, and improvements. When implemented right, technologies are flexible and agile to respond to any change.

Scaling and transforming

Once these six principles are implemented into the digital transformation strategy, the results from the initial use cases will follow. This success should be leveraged to gain momentum and initiate collaboration in the following stages of the strategy. As digital transformation continues, an organization will identify new ways to transform the business digitally. When it comes to scaling, businesses must scale both horizontally and vertically. Scaling horizontally refers to applying similar strategies to more locations while scaling vertically suggests connecting additional technologies.

As already said, digital transformation is without its destination. If an organization implements all six principles, it doesn't imply that digital transformation is completed. With the dynamic environment these organizations operate in, it is impossible to expect nothing will change from the first moments of working on the strategy to obtaining the desired business outcome. There will always be changes pushing organizations to seek answers through a digital transformation, over and over again.

KEY PLAYERS IN DIGITAL TRANSFORMATION STRATEGY

There are three key groups involved in creating a digital transformation strategy in organizations – the communications professional, the technologist, and the leader. They are incredibly valuable for developing a strategy to transform the business and improve the market position. An in-depth digital transformation strategy must include all three groups in the strategy development phase.

COMMUNICATIONS PROFESSIONAL

With every organization, there is always more than just one target audience. Organizations communicate to their employees, customers, prospects, potential employees, investors, etc. Business value can be explained through communication. When something changes in the organization, communication becomes even more important. With digital transformation, there are two stakeholder groups – internal and external.

Internal stakeholders are employees, who need persuasion that the changed state is better than the last one. A skilled communications professional helps organizations understand their employees and empathize with their need to fight change. Without understanding the employee perspective, organizations cannot convince them that digital transformation leads to something better, and a communications professional can do both of these things.

Communications professionals understand that people are not afraid of the change itself, they fear the unknown, so their role is to demystify the unknown by answering all questions and concerns employees might have until they possess a clearer understanding of digital transformation. Employees don't have to have the same amount of information as the business leaders, but they need to have enough to feel like they are not stepping into the unknown.

The communication professional takes their guidelines from the digital transformation strategy. There, they find what information is necessary to communicate to other employees and at which stage. Without the communications professional understanding the vision, it will be difficult to have all other employees on board. This also implies that the digital transformation strategy must be clear and detailed enough so all stakeholders can act accordingly upon it.

TECHNOLOGIST

If an organization has excellent technologists, they will probably not be even aware of their presence. The job of every technology department is to ensure all threats and issues are resolved for the business to run smoothly. Until their business is affected, most business leaders forget about software developers, app developers, systems integrators, network engineers, and data professionals. Yet, this doesn't diminish the role of a technologist in digital transformation, which is one of the key players to success.

The communications professionals will ask technologists to help them with actions that might result in data leaks or any other cyber threat that might harm business in any way. On the other hand, technologists should be involved during the entire digital transformation implementation, not just when things go wrong and they are called in to solve the issue. With proactivity and being present during the entire transformation, technologists would help organizations by preventing any potential problem and maintaining all systems at all times.

To maximize the knowledge, skills, and experience of a technologist, the organization must first change its mindset. They are not the maintenance professionals called when something happens, but valuable stakeholders in digital transformation. Leaders need technologists during experimentation and both sides need to change how they approach experimentation in general. Technologists shouldn't hide mistakes anymore because they fear the consequences. Transparency is crucial when implementing a digital initiative successfully. Mistakes that technologists make during experimentation are expected and could still lead to a positive outcome.

Business leaders should avoid putting too much pressure on their technologists during experimentation. These high expectations pressure technologists into hiding their mistakes and affecting the progress of the digital initiative. Correcting the mindset from both ends is essential if businesses want to thrive. Even the best technologists cannot digitally transform a business if that mindset stays the same.

BUSINESS LEADER

Besides the communications professional and technologist, the business leader is also crucial when developing and implementing a digital transformation strategy. Business leaders are in charge of creating a data-driven strategy that brings clarity to everyone and helps them understand their role in digital transformation. However, creating a data-driven strategy is everything but easy. Such a strategy involves understanding where the emerging business opportunities will come from and allocating the resources to exploit the opportunity when the moment is right.

Yet, the digital transformation strategy doesn't end with just detecting an opportunity when it arises. An organization must fit the opportunity into the wider picture, determining the way to displace business-as-usual and restructure the entire business to serve the customer more efficiently. As new digital business models appear, organizations are constantly faced with the fact their existing business silos are not adequate to serve the customers. That is why business leaders need to focus on building teams based on customer tasks, which requires a lot of agility.

The responsibility of every business leader is to ensure that customers are getting their needs met at different touchpoints of the organization. Customers can interact with organizations through various communication channels, so a leader must enhance these interactions and maintain satisfied, happy customers.

A digital transformation strategy will not succeed if these three roles are not deeply involved in creating and delivering it. A transformation change is challenging if these three roles are not working together towards the same objective. As each of them brings a unique skill set, they should all participate in developing a digital transformation strategy because the business will suffer without them. In smaller organizations, one person will often take on more than just one role, which can work as long as the person is aware they have two major responsibilities to take care of.

EXAMPLES OF DIGITAL TRANSFORMATION STRATEGIES

There are many different strategies organizations can opt for when digitally transforming their business. An organization can create its digital transformation strategy without copying existing examples. Each organization has its challenges and issues they aim to overcome with digital transformation. In a way, there is no good or bad strategy *per se*. It depends on all factors mentioned in this chapter – components, steps, principles, and key players. If all these factors are covered and implemented well in the strategy, it will be easier to transform a business, regardless of its ultimate objective. If an organization lacks any of these factors, it will struggle and face more challenges during the implementation of the strategy than it can take.

Examples of a digital transformation strategy serve organizations as recommendations instead of rules that need to be followed blindly. This type of strategy requires creative solutions that will improve the ways a business operates and interacts with its customers by making the most of all digital tools available.

LEVERAGING IN-DEMAND DIGITAL TECHNOLOGIES

Investing, leveraging, and utilizing digital technologies that show positive results will bring the organization closer to people. Determining which digital technology is the right choice to capture the attention of potential and existing customers and facilitate how all stakeholders within the organization work and interact with each other is complex.

An organization will typically have several types of audience, and they all will not necessarily have the same expectations from the organization. So, before leveraging any digital technologies, business leaders must be aware of how chosen technologies serve all stakeholders. This often requires testing and modifications of digital technologies. Also, it is typically the moment where business leaders seek help from technologists.

With the right digital technology, businesses can build value and enhance efficiency on their existing business models. Companies need to prioritize all strategies empowering digital technologies and be aware of what to use them for, how they function, and how to use big data and information. Ultimately, strategies that leverage in-demand digital technologies boost the organization's productivity as well as the quality of its products or services.

There are three types of digital technologies that have proven to bring enormous benefits to any business type – project management, time tracking, and social media management tools. Each improves efficiency on all organizational levels when implemented the right way.

Project management tools

A project management tool can help organizations have more productive and cooperative teams by having a clear overview of how everyone is investing their time and effort. A project management tool improves transparency because every task has a responsible person, deadlines, and comments.

Time tracking tools

Time tracking tools help organizations measure the performance of each member, whether they are working on smaller tasks or bigger projects. Because of its deeper insight on employee performance, time tracking tools can help team leaders address issues on time and improve the performance of a person or entire team to ensure the objective is reached as predicted.

Social media management tools

With automated digital tools that simplify managing of all organization's social media accounts, social media specialists will be able to create content in advance and schedule them within the tool. This leaves them with more time to focus on analyzing social media data, which is tremendously valuable to other marketing and sales strategies.

ENHANCING CYBERSECURITY OF USED TECHNOLOGIES

Even though we live in a digital era, security is considered important but not a priority. With organizations treating security as just another area that needs to be taken care of when working on digital initiatives, it is not easy to transform a business digitally. Such an approach often leads to breaches in the system. Lack of cybersecurity is a serious threat to any organization as the number of viruses and cyber incidents continue growing, and technologists and leaders cannot ignore it because the consequences can be disastrous for the business.

Companies that managed to overcome cyber challenges by implementing digital initiatives recommend hiring Chief Information Security Officer (CISO). Many organizations are still unaware of cyber threats, so it is difficult to justify allocating the budget for another tech role. However, this role is crucial for ensuring all valuable information and data within the organization are kept and secured appropriately. Otherwise, the security breach that could occur might disrupt numerous business operations in the organization.

When talking about security breaches, each organization is responsible for protecting the information and data they possess. Besides hiring a Chief Information Security Officer, many other actions might improve cybersecurity across used technologies in the organization.

Changing passwords regularly

All employees should be educated and reminded to change their passwords. Also, they should always create strong passwords each time they replace the old with the new one. To go one step further in protecting information and data, organizations are using passwords manager tools, which simplify managing and sharing passwords without compromising them.

Ensuring all connections have VPNs

Every organization should implement Virtual Private Networks or VPN connections between their office locations. VPN connections protect the organization's network, which is crucial for those connecting through public Wi-Fi services. That is why all connections used by the organization must be secured by VPNs.

Installing Ad Blocker in computer browser

Ad Blocker is one of the trusted cybersecurity features in the computer browser as it protects employees while browsing the internet. This significantly decreases the chances of bugs entering the computer and affecting the cyber health of the organization's computers and systems.

Using two-factor authentication

All employees should use Two-Factor Authentication to prevent any hacking incidents. Many organizations have witnessed other entities trying to log into work emails to access sensitive information. With Two-Factor Authentication, organizations protect vulnerable data and employees from security breaches.

ALLOCATING THE BUDGET R&D OF DIGITAL TRANSFORMATION SYSTEMS

Expanding the budget for research and development of systems will result in long-term sustainable growth, and there are no valid reasons for organizations to continue avoiding it. Although many business leaders expect that their revenues will grow after the digital transformation adaption, there is still a huge discrepancy between these beliefs and increasing the budget to maximize the potential of digital transformation systems.

Technology cannot improve on its own, but it requires exploration and testing to determine how it could contribute better to the whole organization. With profound research and development, IT systems can be maximized and optimized. With an increased budget, organizations can leverage their technologies and take their business to a higher level. Not to mention that optimized digital transformation systems ultimately result in enhanced business operations in general.

Giving autonomy to the R&D team

A research and development team will not be able to do their job if someone else makes the decisions for them. These experts have the needed knowledge and skills to make their decisions and detect potential opportunities that will lead to positive business results. Including them in conversations about the market's performance will provide them with valuable information to do even better.

COOPERATING WITH R&D

As much as the R&D professionals need autonomy to make their decision, they still need the help of other teams and other business operations. In one organization exists the same culture, so sharing information among different departments can help maintain high-quality products while also improving efficiency.

Developing user-friendly systems

The user-friendly aspect is not mentioned enough in digital transformation. Every transformation aims to enhance certain business areas, and what makes it successful are the people using these technologies and systems. Organizations build digital systems to support and perform daily tasks and replace humans with machines. This allows the human workforce to focus on more complex tasks and functions that will be more beneficial for the organization.

However, digital systems need to be user-friendly. The purpose of having a certain digital technology is to allow the human workforce to move onto other tasks and projects, but complicated and confusing systems won't bring any benefits. It will bring more issues and challenges for employees who will need to manage these systems, leading to increased costs and time spent.

Each digital system an organization decides to implement should be efficient, simple to navigate, install, and remove, and without the need to use third-party software. These systems must be effective when handling errors and follow rules or standards set by the organization. The more user-friendly the system, the more successful its implementation will be, followed by numerous benefits on a much higher level.

DIGITAL TRANSFORMATION ROADMAP

Many organizations will talk about a digital transformation roadmap, but only a few will know how to deploy one and get the most out of it. Establishing an efficient process addressing all crucial needs of a unique organizational environment allows businesses to find what works well for them and to create a strategy to make it real.

A digital transformation roadmap is a well-thought plan that moves an organization from here (point A) to there (point B). In the current situation, organizations are using their existing processes, but to get to where they want to be, they will need to start using new digital processes. A digital transformation roadmap helps organizations define and manage the digital initiative. It provides them with a structure to migrate from one tool to another, including technology, people, processes, and other resources required for a successful transformation.

However, just having a digital transformation roadmap doesn't guarantee it will be completed successfully. Organizations use different methods to ensure their roadmap is being successfully deployed, and one of the most efficient is turning employees into change ambassadors. Organizations need their employees to be on board with the changes they want to implement because they will be directly affected by them.

Even if the digital initiative is to reduce the human factor by replacing it with machines, employees will still need to be involved to a certain point. That is why their opinion about digital initiatives is crucial for success. To turn employees into change ambassadors, department heads or senior management should first understand the purpose and value of the change, so they can go and inform their teams about it.

Also, setting a faster transformation pace and sharing quick wins in the first phase of the initiative improves the company morale. The first few months are essential because they will set the tone for the rest of the digital transformation journey. Because there will be many new things happening in a short period, all employees will be observing, and showing them how it all works as planned is the way to motivate them to participate even more. Employees participating in the first victories should communicate and celebrate with the rest of the company.

When using change management tools, staying on track with the digital transformation roadmap and making the new technology implementation even more efficient becomes easier. Many tools can support different aspects of the roadmap. One of them is the digital adoption platform (DAP), which empowers users to start maximizing new tools and processes quickly with in-app walkthroughs, contextual guidance, smart tips, task lists, etc.

Feedback-gathering tools are also an excellent way of supporting the roadmap success because it motivates employees to engage, simplify sharing ideas between them, and gather insights through the process. When it comes to new information, the best way for employees to learn them is by creating knowledge wikis that will serve as a central point for all training materials. Of course, many other methods might facilitate how employees reach digital change and the choice will depend on the needs of the organization and its employees.

TEN-STEP PROCESS TO BUILD A DIGITAL TRANSFORMATION ROADMAP

To build a digital transformation roadmap set for success, organizations should follow a straightforward ten-step process. From gathering and analyzing data and prioritizing issues to creating a culture passionate about digital transformation, an organization will need to create a comprehensive roadmap that calculates all potential opportunities and risks and prepares itself to reach them.

Gathering and analyzing data

Before making any changes, organizations will need to assess their starting point. Which of the existing processes or systems is currently working well? Where can these processes or systems be improved? Which specific changes does the organization expect to see from digital transformation? Gathering and analyzing data before starting with a transformation provides the organization with the foundation to measure how effective is the initiative. For instance, if the objective of digital transformation is to improve team productivity, those implementing the changes will need to know how much time the team members spend on particular tasks before the new tools or processes were introduced.

Once the transformation process is completed, measuring and comparing outputs will demonstrate whether the initiative was successful or not, ensuring to have the right data and insight to set a stable foundation for the rest of the organization's transformation path. This is why it is essential to gather whatever information is missing before getting started with the digital initiative as this helps tackle all the digital transformation challenges an organization might face throughout the digital transformation journey.

Getting senior management involved

When first looking at it, digital transformation can seem like a simple change in technology or process. To succeed, digital transformation requires a complete cultural change. Implementing a new system within the organization will require training and often a learning curve, which implies that the entire team needs to be committed to achieving the objective of the digital initiative.

However, senior management is mostly in charge of showing the way when changes occur in the organization. If senior management is not committed to making the initiative work, all efforts to change or improve something will fail. So, for a digital transformation to succeed, senior leadership support is needed. This means that those who are aware of the value that lies in the particular change will need to gather the right information to convince the senior management on why this change is essential for the business.

To connect with the senior management, those who seek to drive changes in the business should relate digital transformation to the organization's goals and values. Explaining how undergoing digital transformation will bring the organization closer to achieving its major objectives will have an impact on the senior management. Also, demonstrating to them how this change will help the organization better align with its values will show them that the change is not the additional element to the organization, but its evolution.

Focusing on ROIs will also prove to senior management that the investment required from them will have positive results for the business. Another thing that all senior management members love to hear is numbers. As much as digital transformation is a complex process that often cannot be reduced to just numbers, presenting a case supported by numbers will have a greater impact.

Quantifying and prioritizing issues

In the initial digital transformation meetings, a lot of issues will probably come up. Keeping in mind that these issues can be solved and that they will not end digital transformation is crucial. Many times, those who advocate for change will feel discouraged because nobody will understand their enthusiasm and motivation for that change to happen, but it takes time to convince people that embracing change brings benefits.

Instead of getting discouraged, those seeking change should identify these issues and note them down. Then, they should take a look at the entire list of issues brought by other employees and start classifying them. Once categorized, these issues will need to be presented with a solution the next time the change seekers discuss digital transformation with the management of the organization.

One of the ways to present these issues is by creating a high impact vs effort chart. This chart allows detecting whether high-impact opportunities have issues associated with them. Oftentimes, organizations will be overwhelmed with the potential benefits of their digital initiative that they will fail to see all obstacles around opportunities. Creating a plan with solutions for each issue that might arise is the only path to digital transformation success. These solutions must be detailed and they should mention all resources needed for each issue to be resolved.

Detecting potential issues on time allows organizations to prepare for the obstacles they will find on the way. For instance, if an organization is seeking to start gathering a large volume of data, identifying the need for hiring a cybersecurity expert will reduce the chance of data breach or any customer data violation in the future.

Setting goals

An organization must set clear, realistic, and measurable goals to drive the initiative. Setting these goals helps organizations set themselves up for success and achieve the ROI much faster than without them. To set goals properly, the organization will need to determine the ultimate goal of their digital transformation process, also known as Point B in the roadmap. Most organizations avoid setting ultimate goals because it feels far away, too ambitious, and difficult to achieve, but knowing what is on the finish lines helps discover all the obstacles that need to be overcome to get to the goal.

Most of the time, the ultimate goal of digital transformation will require a phased approach, followed by smaller, more achievable goals. For instance, if the ultimate goal of one organization's digital initiative is to increase team productivity, a smaller goal can be to replace tools proven to be time-consuming and challenged to manage with simpler alternatives.

When establishing goals for digital transformation, organizations should follow the S.M.A.R.T. structure:

- **Specific:** Determining which goal the organization seeks to achieve.
- **Measurable:** Deciding on the metrics that will indicate that the goal was achieved.
- **Achievable:** Considering whether the organization has the capacity the achieve the goal.
- **Relevant:** Identifying the way the goal aligns with the organization's business plan.
- **Timely:** Determine the deadline for the goal.

Establishing accountability

An organization cannot begin or complete a digital initiative without its employees. As much as employees must be informed about digital transformation from their managers, they should also be given responsibility for the projects they will be working on. When the digital transformation strategy is approved, it will need to be broken down into smaller projects that can be given to certain employees or teams in the organization. They will become set owners who will be accountable for completing the project they were assigned to.

That said, giving a digital transformation project to employees is not the same as assigning them regular work they are used to. For most employees within the organization, digital transformation will be a completely new area and they might feel confused or lost. That is why it is important to provide them with all information and guidance they need. Some organizations decide to employ one person to serve as a center of information for all those working on digital transformation projects, while others increase the availability of managers to employees to ensure there is enough time to cover all questions and track the progress.

Ensuring employees have all they need to execute their project successfully is essential, but delegating authority to certain areas will get the project done quicker and more efficiently. Managers are aware of the capacity and skill set of each team member, so allocating work to certain employees boosts productivity and team collaboration. Also, this provides managers with more bandwidth to handle other issues with a higher priority.

Allocating time and budget

Digital transformation should be approached as a series of projects the organization plans on the undertaking, meaning it needs to be considered in terms of time to money to excel. Organizations should plan regular meetings to go through the issues that might affect the delivery of these projects, but also think about how much time and money each of these projects require.

Each owner of a project should have a budget that will allow them to complete it successfully. While allocating the budget, it is important to also discuss the impact on business resources to keep everyone on track. If any changes in budget or time resources appear, everyone working on the project should be informed on time. Regarding deadlines, managers should decide what is the time required for the project to be completed successfully. Input from those working on it here is essential as they will have the best idea of the time needed to complete the project.

Taking a phased approach

As digital transformation will be divided into smaller projects, each project should have a timeline and related subtasks. Many organizations will start implementing all of these mini-projects at the same time, which leads to overwhelming the team and poor performance. Instead, business leaders should execute, review and create a retrospective for each task. This will help business leaders to look back to see the progress and detect ways to work quicker and more efficiently.

By approaching the digital transformation process in phases, it is much easier to understand whether the performance was good or bad, gather feedback from employees and make adjustments where it makes sense. Testing continuously and planning globally allows organizations to see valuable benefits from completing these smaller tasks, while also ensuring they don't fail to complete other mini-projects within their digital transformation process.

Creating a culture passionate about digital transformation

The importance of getting everyone within the organization committed and excited about digital transformation is essential for its success. If a team doesn't see the value of new processes and technology that will be implemented, it will be quite challenging to get them involved.

When employees are informed and included in the discussions about digital initiatives, they will be more interested to see how it works. Everyone in the company should know why there is a need for change and how this change will enhance efficiency and productivity. Also, employees should be encouraged to ask questions and raise their concerns. This will benefit the business leaders as well because once the transformation begins, everyone will be well informed and will not interrupt the process in any way.

However, it is vital to keep in mind that people are hesitant to change. Organizations must allocate a certain time for adjusting to the proposed plan of action. Expecting that everyone is on board with the digital initiative from the first day will only lead to confusion and frustration between everyone in the organization.

Investing in agile project management

Agile project management focuses on building a collaborative environment to implement, test, and respond to such changes. Agile project management provides organizations with the chance to continuously revisit their plan and adjust according to the needs and concerns of their employees.

It also improves the success rate for all transformation efforts by dividing the project into sprints. This ensures better tracking of how each bit of change impacts the organization, including getting teams' feedback on what is working and what isn't, to improve the implementation process on time. Agile project management can be best described as a co-pilot in the passenger seat on a road trip to the ultimate goal of digital transformation.

Seeking post-implementation feedback

Once digital transformation has begun and everyone knows what they will be working on, it's important to communicate with those implementing the changes. Employees who are directly working on these tasks will know best how was something implemented, what issues might appear, and how to find alternative solutions. However, their feedback is not only valuable during the initial phase of digital transformation.

Organizations that have implemented the digital changes should seek post-implementation feedback from everyone in the company. After all, everyone should be affected by the implemented changes, so it is only natural to seek their feedback about it. Once the tasks have been implemented and transformation is seeing its end, the organizations should start gathering feedback to measure how successful the implementation was.

The best way to do that is by surveying the people and partners impacted by these changes and analyzing the responses to ensure that all changes are optimized on time and employees and customers are maximizing all the benefits from digital transformation. Also, asking those involved about what they would do differently and redefining the process to improve where possible will ensure that digital transformation was successful.

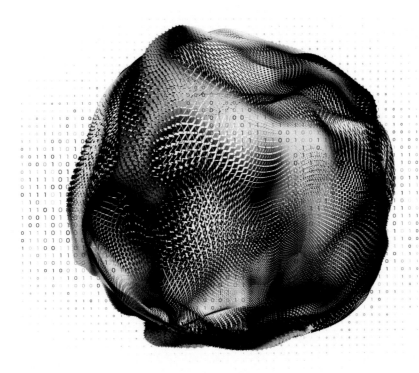

Chapter 5

Data-driven society

With more users on social media platforms and other interactive platforms on the Internet, people are creating more personal data every day. Humans are not the only ones adding this data because the sensors in the environment around us are also picking up data constantly. From Google, Apple, Amazon, and Facebook to other tools and platforms people use for professional and private purposes, these companies focus on gathering and managing our data. The entire process of collecting personal information is difficult to understand from the user's point of view, so many are not even bothering to learn about it (Figure 5.1).

There are numerous forms of data generated continuously, including personal data. Once gathered, these companies will store and manage this data somewhere. The purpose of data might differ from one organization to another. For instance, one company might use it for their one-on-one marketing strategy, ensuring their services match their customers' preferences and personal interests. Another company might use this data to create a new product that will be precisely what their customers need.

As this movement continues growing and more interactive online places appear, it is safe to say that we all collectively are approaching the age of the data-driven society. The connection between data and the physical world is so sophisticated that it is often challenging to notice it, yet it produces a new value for organizations across all industries. Data tremendously affects how we live our lives and choose between different options, even when we are unaware.

When a person is looking for a suggestion on where to eat, they will often depend on the restaurant recommendations from a search engine. The same goes for online shopping or any other online activity where you seek a certain piece of information. As much as these examples might seem obvious, they are confirmations that we have already entered the age of data-driven society. As our lives are becoming intertwined with data, the quality and speed of the decision-making process should also improve. Potentially, this might enhance the quality of life, but humans will need to learn how to use their data on their own.

DOI: 10.1201/9781003305163-5

Figure 5.1 Data-driven society.

All organizations, which are also data holders, have the immense responsibility of ensuring their customer data is secured and managing them in a way that serves the customers. That said, it's also crucial for customers to be aware of giving correct personal data, especially in certain situations.

Not to mention that technological drivers like the Internet of Things, blockchains, and artificial intelligence are interrupting human lives like no other technology before, so it is essential to regulate this future data-driven society. As most existing national legal instruments have already started tackling the challenges put in front of regulatory institutions to protect personal data from any threat that might occur in cyberspace, they need to revise these legal instruments and update them instead of creating new ones each time a new challenge appears.

Today, society relies on data-driven services to enhance the quality of citizens' lives. This data-driven culture could enrich transport, energy, and health, to name a few areas. The wider it becomes, the bigger the need for a common regulatory umbrella aiming to maintain a harmonized vision of cybersecurity.

But, what does data-driven society mean for organizations that are data holders as well? If an organization gathers data, it can undoubtedly help with marketing and sales strategies, customer service, and product launches. To make any of these business objectives successful, an organization will need access to valuable data. However, when this data is disorganized and not properly handled, it will be almost impossible to get any value from it.

Organizations need quality tools to help them with data visualization, data analytics, and how this data ingests. With an overwhelming volume of gathered data, organizations are turning to technologies like artificial intelligence, machine learning, and data warehousing to turn it into a constantly-enhancing driver of strategic decisions and actions. But, getting there is a long, frustrating process, and it requires deep knowledge on how to use data to benefit the organization it is gathering data in the first place.

DATA INGESTION

Before an organization can do any type of data analytics or the application of cognitive systems, it will need to ingest the data it gathered. Organizations can take a few data ingestion points and the most comprehensive one is working with an intelligent data warehouse because organizations can integrate that data later with a range of systems, such as data lakes, artificial intelligence, machine learning engines, etc.

Data warehouses used to be based on a rigid data model, including only structured data. New data warehouses can ingest even unstructured data and conclude data models and schemas on the go. When the cloud appeared, it helped create the next generation of data ingestion points and locations where data warehouses come in. A cloud-based data warehouse implied that the entire design process should be more lightweight. Of course, identifying data sources and user needs is still crucial, but the data ingestion happens with just a click of a button. More importantly, organizations can explore and transform this data while in the data warehouse.

This specific workflow is called Extract-Load-Transform or ELT. It provides mighty flexibility for data engineers and analysts as they will not have to come up with the entire process, from the data model to the OLAB cube structure. The ELT workflow allows them to define the process when new data pours in the data warehouse and new user needs appear. These cloud-based data warehouses show a more efficient way to extract information from data and analyze it. More importantly, they are accessible and effective not only for larger organizations but also for small and medium ones. These companies no longer have to spend millions of dollars and wait for months to have their monolithic data warehouse set up. With cloud-based data warehouses, organizations can set it all up in just a few days and with several hundred dollars per month. Time and money are crucial factors for any business, and cloud-based data warehouses save both.

All this new technology allows organizations to maximize the data potential by:

- Exploring and better understanding source data,
- Preparing, optimizing, and delivering data solutions,
- Integrating with data visualization to allow deeper visibility into market patterns,
- Generating data models for a range of use-cases,
- Deploying data-driven solutions quicker,
- Reacting proactively to market trends due to increased insights.

Simply, organizations that work with data warehouses will benefit from data agility, which allows changes and improvements each time the market shifts and changes. This level of agility is crucial for creating competitive advantages and staying ahead of the competition. Every organization working with numerous data sources and struggling to find the value in its data should start by rethinking how it ingests this information and where a data warehouse seems like an efficient solution.

DATA-DRIVEN SOCIETY IN DIGITAL TRANSFORMATION

There is no doubt anymore about the incredible value data holds for organizations. Workplaces are becoming more dynamic, and to be ahead of all changes that affect business, organizations must adopt and implement the needed cultural changes. Digital transformation is impossible without data. The initial step towards digital transformation is learning how to manage data appropriately and eliminating all obstacles preventing the organization from using it. Although more business leaders are becoming aware of data being a valuable asset in their organizations, there is still a lack of understanding of its transformative potential and how to utilize it the right way.

Before data, leaders relied on their gut instincts, an ability to decide based on their intuition, but this quality has become completely undesirable in modern business settings. Instead of relying on intuition and crossing fingers their hunch will work, business leaders need to switch their mindset to data management and analysis. At the core of every data is the potential for transformation that might enhance one or more business aspects by optimizing processes and facilitating certain tasks and projects for employees.

Not to say that organizations that have yet to learn how to ingest data need to implement advanced data analytics and catch up with the rest. Besides being impossible, it will cause more damage than good to these organizations. Like digital transformation is a process that requires time to mature, learning how to work with data will not happen overnight. Whether organizations seek to satisfy their customers or improve their workflow with data, they will need to go through the entire process step by step as its complexity allow no alternative. A clear data strategy with modest actions, perfectly aligned with an ultimate objective, is a way to successfully manage data and transform them into a valuable asset for the organization.

In such a data-driven society, all business decisions are made with data. From simple financial figures to advanced analytics results, organizations decide on their future based on data they gather from various data sources. In a data-driven society, data becomes the primary source of insights in each department of an organization. Data is a cornerstone of each department's strategy and the overall business strategy that drives the organization forward. The concept of being data-driven focuses on numbers, but the way these numbers are interpreted is where the value lies for these data holders.

In organizations, data is used to empower everyone to make more informed decisions, enhance initiatives, and improve the competitive advantages of the organization. Also, the objective is to build a collaborative culture among all employees of the organization to allow data to be that cornerstone for them.

Finally, the objective is to build a collaborative culture that involves all organization members to ensure that data is the basis for all business decisions. Data should help everyone make more informed decisions, from the data owner to the data analyst and everyone else who will use it. To make the most of it, organizations will need to develop new, data-driven applications, detect data patterns, and experiment with analytics platforms to determine what works in all these processes. A data-driven society is enabled by access to data, data quality management, methodological knowledge for data analytics, and technologies that allow them preparation and analysis.

Many businesses fall into the same trap with data and digitized projects. They see this initiative as an opportunity to upgrade technology, assuming it will enhance workflows and productivity simply by incorporating new systems and applications. Although it will have a certain impact, such thinking lacks critical strategic focus. When trying to maximize the potential of data, these organizations need to have a clear goal determined and set of actions that will help them achieve it.

A digital transformation initiative will typically involve migrating from offline, manual processes to online, digital ones. In a sense, this means that all digital transformation projects are also data projects because they involve digitizing data, content, and any relevant information within the organization. On the other hand, each organization determined to implement any digital initiative will need to consider how its strategy will lead the organization to the desired business outcomes, but most of them will overlook the implications of such transformation on data.

For digital transformation to be successful, it must work in two ways. The gathering and use of data have a tremendous impact on the digital plan, and the way this data is managed impacts how successful digital transformation will be. That said, it is pretty evident why data and digital strategies need to be aligned.

For organizations, data has become one of the crucial elements. In this digital era, data is unparalleled to anything else. Only when organizations have access to the right data, the platforms can create personalized and customized moments of realization for their audience. In the digital transformation process, these platforms are called digital experience platforms or DXPs.

A data-driven organization can use digital experience platforms for:

- Identifying immediate customer experience needs to scale the business,
- Organizing around the customer journey by making data and insight actionable,
- Training teams on data and analytics, leading to building a data-driven mindset.

To create data-driven leadership, organizations need to start with efficient data leadership. Once the culture of data-driven leadership has begun with its implementation, business leaders will need to monitor how it is manifesting and provide their teams with the right tools.

To ensure an evolving data-driven culture in an organization, leaders will need to take a few steps. First, an organization will need to develop a clear vision of success, involving data implementation into every business aspect. Second, they will need to implement a data-driven mindset by enabling employees to access data easily and create a culture shift within the entire organization. Third, data will need to be maintained clean and clear at all times. Fourth, an organization will need to build agile teams rather than just focusing on implementing the right tools. Finally, organizations should consider implementing a reward system to encourage healthy competition between the employees.

BIG DATA

We cannot talk about data and how it impacts digital transformation without mentioning big data. For organizations that don't understand big data and how to leverage its power, this concept can seem a bit overwhelming. Big data is the large volume of data collected by organizations daily. It is characterized by its enormous volume, variety, and complexity, which makes it challenging to process using conventional data management practices. Due to that, big data needs fresh and innovative data processing methods like data analytics (Figure 5.2).

Data analytics processes big data and extracts valuable information from it. Organizations use this information to make marketing, sales, and business decisions while going through digital transformation. With digital transformation, organizations can embrace change and remain competitive in their industry, but the real value of big data in digital transformation comes from organizations' ability to combine both to allow digitization and automation of organizational operations. Both digitization and automation result in improved efficiency, boosted innovation, and new business models.

Big data analytics enables organizations to have detailed information about specific customer groups. This information can come from the actions

Figure 5.2 Digital world.

customers take when on the organization's website, products or services they buy, how often they buy, and if they will buy the same product again. By using all this granular information, organizations can implement changes to meet the needs of their customers while determining the way to meet these needs. To complete their digital transformation, organizations should adopt both big data and data analytics.

When using big data and analytics in business, data presents a separate challenge on its own. Many organizations gather a lot more data than needed or unnecessary types of data. As the volume and type of data an organization gather grows, the complexity of data analytics also increases. Therefore, organizations should narrow down the data types which would be most valuable to them. This would reduce the data volume they gather in the entire process.

So, before an organization starts collecting data, it should identify the biggest short-term and long-term challenges. This list of challenges will help an organization to break down the data it gathers to useful insights that can be used to make more informed decisions in business and drive growth. With more organizations becoming aware of the data value, the demand for talent with data analytics skills continues growing. This is also one of the main reasons why data science and data analytics skills are some of the highest demanding roles in the field.

When setting the objectives data helps to achieve, organizations need to be as specific as possible. Objectives such as improving the bottom line are not specific enough. An organization should focus on finding answers to how to do it and what to do instead. For instance, business leaders can set specific objectives such as retaining customers and reducing operational costs, as both of them improve the bottom line.

Once the organization has identified its objectives, it should focus on gathering data sets that help it meet its objectives. If the objective is gaining new customers, the organization can focus on its data from social media platforms and other sales channels because such data often shares valuable information for customer acquisition strategies.

However, big data and data analytics can obstruct digital transformation if data is not managed properly. Having access to a large volume of data means nothing if the organization is unable to organize and manage this data to facilitate its usage. To maximize the potential of advanced analytics and machine learning models, the data needs to be trustworthy. This is why it is essential to gather the right data and manage it efficiently. Ensuring the data is trustworthy helps organizations enjoy more benefits when using this data.

While organizations have this urging need to hire a data analyst to bring sense to all this data they continuously collect, small organizations collecting a small volume of data or those without enough resources will not be able to follow this path. They can use a range of tools and platforms that allow business leaders to gather data, segment it into data sets, manipulate and organize them in a way that the entire organization can evaluate and understand them. Also, these tools help small businesses to track the impact of the decisions they made by utilizing data and current industry trends and projections.

When an organization has the tools to analyze data sets, with or without a data analyst, it can start taking steps to digital transformation. Such digital transformation, based on quality data, can then be used to gain a competitive edge in any industry.

WAYS BIG DATA REVEALS DIGITAL TRANSFORMATION OPPORTUNITIES

When managed properly, big data brings to the spotlight neglected, dark corners of the organization. A large volume of well-managed data can deliver a better comprehension of operations, customers, and markets if implemented within an analytics or artificial intelligence program. For a digital transformation to succeed, organizations need large amounts of data and quality management.

On its own, big data is completely useless unless the organization has a digital transformation strategy to make use of it. Big data ensures that organizations gather a large volume of data, which ultimately increases the chances for success. When digital transformation and big data converge, a change of real value for the organization becomes possible. As the number of IoT devices, smartphones, and wearables grows, the amount of data these devices generate also grows. The combination of all that data, the potential of big data analytics, and digital transformation enables organizations to adjust in almost real-time to their customer needs, but also predict their behavior in the future.

The growing increase of internet-connected devices, evolving data-driven business models, and globally connected business ecosystems require that organizations build a cohesive, modular digital platform, powered by big data. The return on investment that an organization can generate from its investments in the digital platform will greatly depend on its data value extraction potential. When digital technologies focus on deriving maximum value from big data, it can allow technology leaders to build data hubs for accumulated and staging data from various sources. Many big data providers offer pre-built analytics and machine-learning algorithms that these leaders can leverage and implement in their digital transformation initiatives.

However, big data and linked digital transformation efforts must be clearly defined for the organization and the industry. For instance, an organization will need to define whether the goal is to increase revenue with connected products, cut costs with a more optimized platform, or something else. Only when the objective is clearly defined, IT leaders are ready to determine an in-depth strategy for their big data, IoT, and cloud.

Every digital transformation effort should have an objective in mind. Whatever the goal of the digital transformation strategy, all actions that follow need to align with it. More importantly, a digital transformation strategy defines the path and guides all employees in implementing chosen technologies.

Another thing that organizations need to be aware of is losing the business perspective. Too many digital transformation initiatives start within IT departments and fail to succeed because they weren't business relevant. If an effort involving big data is executed on its own, it will seem like a solution that lacks a problem to solve. The most successful digital transformations allow technology, including big data initiatives, the glide path. Big data and digital technologies enable organizations to understand better their customer preferences and behavior to create more personalized experiences.

This introduces a range of insight-based products and services and allows the organizations to combine big data and digital transformation to design new products and services that give them a competitive advantage. Yet, if big data is poorly managed, it might slow down or harm the entire process of digital transformation. That is why organizations need to understand the value that lies in metadata management, data catalogs, data quality, data ownership, and assigned security. All these factors impact more digital transformation than having fewer data to work with than you expected.

Technology leaders being proactive in using big data in aid of digital transformation will be the most successful ones in their industry. They will begin their digital transformation by crafting a strategy for data management. After all, a digital transformation cannot be successful if its data lacks trustworthiness. Organizations that invest in data governance, advanced analytics, and machine learning, will see the most benefits from their overall strategy. The most common benefits that come from such strategies are improved operational efficiency, enhanced customer experiences, and increased revenues.

DATAFICATION

Datafication can best be explained as the transformation of social action into quantified data which allows real-time tracking and predictive analysis. Datafication refers to taking previously invisible activity and converting it into data, which can be monitored, tracked, analyzed, and optimized. Many emerging technologies have enabled numerous new ways to *datafy* daily activities within an organization.

It is a technological trend converting various aspects of organizations into computerized data by utilizing processes that result in data-driven enterprises by giving new forms of value to data. Datafication aims to turn the most basic, daily interactions of humans into a data format, which can later be used for social purposes.

There are already numerous examples of datafication. Social media platforms like Facebook and Instagram collect and monitor data information to market more targeted products and services to their users. Promotions we see on social media are also the result of the monitored data and more importantly, all of this changes our behavior. When seeing these target promotions, we decide to make a purchase, learn more about the advertised company, subscribe to their newsletter, etc. In this example, however, data is being used to redefine the way content is created by datafication instead of informing recommendation systems.

Besides social media, there are other industries where the datafication process is already being used, such as insurance, banking, human resources, hiring, and recruitment, to name a few. Insurance companies use data to update their risk profile management and business models, while human resources use data to detect employee risk/taking profiles, among others.

Maybe the most interesting example of datafication is Netflix, an internet streaming media provider. Many are unaware of the fact that Netflix started as a DVD rental company 20 years ago. Aware of the power of personalization, Netflix was recommended and mailing DVDs to its customers. Yet, this personalization was very limited, allowing Netflix to operate based on data points like the past rental history, the length of time each DVD was held, and basic demographic information. When the company launched a streaming service, the datafication of user behavior started. Now, the company had access to information about user browsing history, points where users press forward/rewind/pause, titles they add to the wish list, etc.

This allowed Netflix to divide its customers into thousands of micro-clusters or taste communities, where each individual can be a part of different taste communities. Netflix uses datafication to better understand its users to provide them with a more personalized experience, from a customized homepage to a personalized Recommended for you list.

Undoubtedly, data is the new gold. With more data, improved algorithms, and enhanced products, organizations can support business model innovation and create more value for their customers. That is why every company that starts with digital transformation should venture on a datafication journey, which involves continuously extracting data from activities and transactions that occur naturally in the organization and establishing data pipelines that allow large amounts of data with high velocity.

Starting to implement datafication in an organization means to be aware of the value that lies in in-depth knowledge of the customers. After all, companies like Amazon and Google are leaders in their industries because they have a profound knowledge of their customers. As humans, we are more social and predictable than we think. As most of our daily activities include smartphones and the rest of the devices, we are leaving behind *digital bread crumbs*, recording our behavior even if we're not aware of it. This digital information about each person can be very valuable to those who know how to use it right. Companies like Amazon and Google do, and they have analyzed millions and millions of customer data to understand the market and deliver exactly what they expect.

Data is everywhere, so the real question is not should an organization gather data, but how they should manage it in the first place. For instance, data can show they tend to go to the same fast-food restaurant once or twice a week, purchase a subscription to many platforms (e.g., Netflix, Amazon Prime, and Spotify), order a new card, walk about eight kilometers every day, etc. As mentioned above, this data is gold – but only to those who know how to use it.

There are different ways that organizations can utilize the extracted data. For instance, two organizations in the same industry can have access to the same data, but they will come up with different ideas based on that data. One organization might decide to partner up with a popular brand that is preferred by its market, while the other might decide to launch a new highly-personalized product or service that will satisfy all current customer needs.

DATAFICATION VS. DIGITALIZATION

Many organizations still confuse digitalization with datafication. Digitalization involves the conversion of any type of information into a digital format, such as converting photos into JPEG, music into MP3 files, text to HTML, and so on. Digitalization will increase the amount of available data exponentially. Essentially, digitalization refers to capturing human ideas in digital form to be transmitted, manipulated, reused, and analyzed.

Datafication refers to turning analog processes into digital processes and customer touchpoints into digital customer touchpoints. In other words, datafication involves gathering data within the entire organization, from HR data and financial data to sales data, customer data, and social media data. The Internet of Things (IoT), a network of connecting products and devices with a connection to the Internet, is why datafication is possible. All these connected devices allow businesses to analyze all types of processes within their organizations.

There is an endless number of possibilities to datafy an organization. Any device, process, infrastructure, or customer touchpoint can turn into smart because of the sensors connected to the Internet. As the number of connected devices is increasingly growing, datafication has everything it needs to become possible for all organizations. Shortly, all these connected devices and sensors will surely result in smart homes, smart cities, but also smart offices. There are approximately 20 billion connected devices globally, and this number is expected to reach 75 billion by 2025. Furthermore, the prediction is that by 2035, we will interact with a connected device every 18 seconds.

Over the next few years, as the number of connected devices continues growing, the amount of data will continue to grow as well. This creates an environment of data, allowing organizations to truly datafy themselves. Once they are growing the large volume of data, their analytics will also become more valuable and will provide the business leaders with more in-depth insights, which is incredibly beneficial when crafting strategies.

The first step in building the organization that will excel in the era of digitally transformed businesses is to datafy the organization. However, it is crucial to be aware that datafication is not only a technical challenge for organizations. This process touches upon every area within the organization, such as business workflows, data governance, strategy processes, privacy aspects, company culture, and security. All these aspects must be considered when preparing the business for datafication.

There are almost no limits with datafication, as long as the organization complies with regulations protecting its employees and customers. That said, the datafication of personality utilizing websites and applications is not that simple. Growing attention to privacy and security, but also regulations like GDPR, make storing sensitive customer data challenging for organizations.

When an organization starts with datafication, it should start small. Each organization, regardless of its size and resources, should start with processes simple to datafy. Because many organizations start too ambitiously, their datafication initiatives ultimately fail. Once an organization has gained enough experience with datafying its processes, it is ready to focus on more complex areas of the organization.

The technology needed for datafication is IoT devices and smart sensors, which will be used to streamline and enhance existing business processes. Simply put, the datafication of an organization is the first phase of transforming into a data organization, and later to digitally transformed business (Figure 5.3).

Figure 5.3 Meta search.

BIG DATA ANALYTICS

It is estimated that around 2.5 trillion bytes are generated worldwide and stored by public administration and private companies every day. Also, cities across the world are becoming full of sensors gathering different types of feed regarding weather, traffic, telephony, etc. Besides being a data-driven society, we have also become data.

Another concept that organizations should be aware of when talking about data, is big data analytics, a process of clustering all technologies and mathematical development to store, analyze, and cross-reference that data to detect behavioral patterns. A growing number of organizations are focusing on this new paradigm, thinking it provides a more comprehensive view of how customers behave. This data allows them to offer a more personalized experience, regardless of the industry.

Big data analytics involves a variety of digital concepts which are already familiar, such as data lakes, data mining, machine learning, and repositories where raw data is stored before analysis. Data mining extracts data to be analyzed by a human; machine learning identifies these patterns and performs actions correspondingly. One of the examples is Facebook's feed feature. The social platform learns from our interactions and adjusts the information in our feed according to our behavioral patterns. This ensures we as Facebook users are offered more relevant content to our preferences.

That is why artificial intelligence will become more dependent on big data. This is no longer only a matter of efficiency in tools, but they should also perform in real-time, learn from behavioral patterns, and predict them. Many companies and public administrations have already started using the virtues of data analysis to provide a more targeted service. Walmart, a U.S. retail company, stores data from millions of transactions made by its customers every day to predict which products are going to be demanded at which hour, day, week, or occasion.

Smart cities also depend on datafication. Public administrations use data gathered from sensors all over the cities to become more efficient, safer, and provide an overall better quality of life for their citizens. Dubai, one of the technologically most advanced cities, plans to digitalize all government services and make them available through the DubaiNow app. A datafication implementation example is a monitoring system using artificial intelligence for bus drivers in Dubai, which has tremendously reduced traffic accidents typically caused by fatigue.

Another smart city that has progressed significantly with datafication initiatives is Oslo. The Norwegian capital is increasingly focusing on climate change, so the government has started using sensors to control lighting, heating, and cooling in buildings, accountable for around 40% of total energy consumption. Their goal is to cut emissions by 95% by 2030 with building control through sensors and electric vehicle development and technology.

This is only the tip of the iceberg when talking about the potential of data in society. The technology revolution will continue to transform organizations and data will play a key role. There are still some challenges about privacy lurking around data and a certain amount of legislation is required to trigger a change in society, which can be compared to the arrival of electricity. Only this time, instead of electricity, the society, which includes organizations as well, will need to learn how to make the most of the immense amounts of data they produce daily.

Chapter 6

Digital transformation tools

Digital transformation has been a guiding star for many businesses for many years now. However, most of them still fail when implementing and using the digital tools they have at their disposal. It is becoming evident that leaders of tomorrow are not the ones who have the most resources to implement a range of digital transformation tools, but those who know how to use it to benefit the business.

Digital technology is inevitable for modern businesses. The COVID-19 pandemic has strengthened the need for companies to transform digitally across almost every industry. That is why technology tools and platforms became crucial for organizations to keep up with growing customer demands and stay competitive. Most business leaders will say that the improvement in using digital tools within their organization has increased revenue. Not to mention that digitally mature organizations also see more profits than their competitors who have yet to embark on a digital transformation journey.

Then, why are so many organizations still hesitant to utilize these digital transformation tools? Several reasons can explain it best. Many organizations have been using tools during transformation and failed to succeed. Of course, entities outside an organization transforming will not know whether the reason for failure lies in the tools, but it will decrease their enthusiasm for investing in such tools. Also, every organization needs to justify its investments. When implementing tools in an already complex initiative, it is difficult to calculate its return on investment (Figure 6.1).

That said, the change has to happen in the mindset of those on a decision-making level. Many business aspects do not fit the frame of profit and losses, and digital transformation tools are one of them. Which alternative do these organizations have to provide quick responses to constantly changing customer demands?

It is the responsibility of business leaders to help their employees to sharpen their skills and deploy tools that will allow them to work more efficiently in delivering customer resolution on time. There exists a powerful correlation between performance of employees and the capacity of a business. Keeping in mind that most challenges in digital transformation are not technical but human-related, both employers and employees must be aware of why a new

DOI: 10.1201/9781003305163-6

Figure 6.1 Digital transformation tools.

tool is being implemented to maximize its potential. The resistance is what puts an end to a digital transformation, and it has the same effect on tool implementation, regardless of its purpose.

Organizations will spend millions of dollars implementing new technologies and training and onboarding their employees. Besides the financial aspect, employees have to invest time in training and learning how to use these new technologies. All of this negatively affects the ultimate implementation results. Another way to approach tool implementation is for organizations to use on-demand training, blended learning, or a digital adoption platform, a software layered on top of another software, app, or website to simplify proficiency by guiding employees through key tasks and providing context as employees navigate the product.

Similarly, numerous other tools can help organizations solve problems and get the most out of digital transformation. These tools are the only way for businesses to overcome transformation challenges and continue driving the changes efficiently. Organizations will need to invest a significant amount of money into digitization, and they will need to do it quicker than their competitors.

Another issue that these organizations find on their digital transformation path is a wide range of solutions and technologies available, making it difficult to evaluate which ones are the right choices to enhance productivity or operational quality. Often, organizations will lack time and resources to do proper research while also may not be able to digest new technologies and solutions at a given moment.

To maximize the potential of all these emerging technologies, organizations must rethink their business models and value creation of the entire organizational ecosystem. When participating in innovation ecosystems, these organizations can try and test recent technological solutions in safe environments.

Organizations that are already deep in their digital transformation process can implement all these challenging initiatives due to high-performing and innovative digital tools. There are many reasons why each organization looking to digitally transform should consider researching which digital tools might facilitate the entire transformation:

- Optimizes time
- Accelerates time required to market
- Increases return on investment
- Enhances internal and external communication
- Improves customer retention
- Boosts news sales

ESSENTIAL TOOLS TO SUPPORT DIGITAL TRANSFORMATION

Organizations must equip themselves with efficient and innovative digital tools. Those organizations that are successfully transforming their businesses have proven how modern platforms and numerous communication and organizational tools are becoming indispensable parts of digital transformation and business success. A variety of tools can help improve a digital workplace if they are not considered as only expenses. Using the right tools is a long-term investment into staying ahead of your competitors and improving the sustainability of the organization.

Collaborative suites

Out of many tools available globally, one can simply not be overlooked – cloud-based suites. These collaborative suites allow employees to access data and resources anywhere at all times. For organizations seeking to make the most of their digital workplace, collaborative suites are crucial to successful adaptation.

One of the most used cloud-based collaborative suites is Google's G Suite, which includes a range of well-known tools such as Gmail, Drive, Docs, Sheets, and Calendar. By using G Suite, employees can share information in real time and access an entire set of tools for enhanced communication and collaboration.

Another collaborative suite that is often used within organizations is Office 365 by Microsoft. This collaboration suite simplifies communication, information storage, exchange, and business activity management. Tools like Words, Excel, Outlook, Teams, and OneDrive are commonly used productivity tools in different industries and organization types.

Collaborative suites like G Suite and Office 365 provide numerous advantages like managing data, executing critical business functions, organizing activities holistically across the organization, and preparing for the digital era. When it comes to organization types, both small businesses and enterprises can maximize the potential of collaborative suites because of their affordable prices and scalability that allows choosing which features should be used.

Communication Tools

Good communication is essential for any organization to thrive. Communication impacts the overall productivity and performance of employees, making it a crucial success factor for all organizations. Only a few years ago, organizations using communication tools were larger ones, but as the pandemic forced numerous companies to start working remotely, there was an increasing need to learn what benefits to expect from communication tools and how to use them the best.

In email communication, the information is impossible to edit once sent, and it can also be misplaced or distributed incorrectly. Digital communication platforms enable organizations to centralize all relevant information and direct employees to a single source of truth for the organization's communication. Best examples of communication tools are Slack and Microsoft Team, allowing you to create specific workgroups to ensure all employees have access to important information regarding a certain task or project.

This enables organizations to ensure better circulation of information and better involvement of key players in each process. Everyone within the organization can contribute by proposing solutions, providing feedback or answering questions. Encouraging engagement in the organization is a fundamental element in stimulating employees and driving them to achieve goals. Undeniably, emerging digital technologies are transforming the way employees work. Organizations are no longer constrained by a conventional, physical office. Adapting how organizations communicate and support all their employees, including the remote workers, is that crucial step toward being a digitally transformed business.

CRM tools

Customer relationship management systems can play a key role in the success of an organization's digital transformation strategy. To stay abreast, organizations need to keep up with the growing high expectations and dynamic requirements of their customers. A good CRM tool will offer a few benefits to each organization, such as helping businesses understand and detect all customer needs, automating tasks, shortening sales cycles, and increasing retention, to name a few.

Figure 6.2 CRM.

When seeking an adequate CRM system, an organization will need to ensure adequate information and data, and its integration with other existing tools in the company. There are several solutions for organizations that wish to deploy on-premise or in the cloud or based on the information type seeking to gather. Some of the most recommended CRM tools on the market are Salesforce, Hubspot, and Freshdesk, yet the best tool will tremendously depend on the needs of the organization implementing it. For one organization, Salesforce might be the best CRM tool to manage all their customer interactions and data, and for another, it might be too overwhelming and time-consuming to navigate such a tool (Figure 6.2).

Sales and marketing teams will only be able to make the most of all their processes if they have adequate tools at their disposal. This is why business leaders should always follow new updates in technology, including artificial intelligence solutions as they provide valuable insights and allow organizations to better understand their target market.

Based on the data from an organization's CRM, it is easy to develop a successful go-to-market strategy to meet customer expectations at all times. More importantly, organizations can learn to anticipate how to win sales, prevent customer churn, and stay on top of competitors. There are numerous CRM tools, with some being more detailed and advanced than others. Besides Salesforce, Hubspot, and Freshdesk, other CRM solutions to consider are Zoho CRM, Zendesk, Pipedrive, Insightly, and Apptivo.

Project management tools

Organizations that have successful project management are more efficient, quicker to market, and ahead of their competitors. However, each business will need to choose the right project management tool for its employees, and then ensure that the chosen tool is adopted across the organization and used in the best way possible.

With a project management tool, organizations can help their teams create and schedule different tasks, stick to project timelines, track project progress, improve efficiency and responsiveness, and offer a holistic view for team members and team leaders. Having a quality project management tool results in having more productive employees, successful projects, and satisfied customers.

To encourage the adoption of tools aimed to improve project management within the organization, those implementing them can set up notifications or email alerts when a change occurs in the project status, an employee completes a task, or a concern is raised. Sharing documents with project management tools allows team leaders to address challenging workflows and tasks with multiple people without needing to organize long meetings (Figure 6.3).

Among the most recommended project management solutions are Asana, Jira, Trello, and Monday. Although they all intend to enhance project management in one organization, each of them is unique, so it's best to always compare before implementing any of these solutions. These tools can easily be implemented to any business, regardless of its size, but organizations must consider how a certain tool fits their overall digital workplace, collaborative suite, and other relevant factors.

Other digital transformation tools can serve for digital transformation purposes. For instance, recruitment tools might help an organization automate its hiring and recruitment processes and save time and resources it would usually spend when seeking new talent. Organizations can also implement payroll management or digital accounting tools that simplify accounting and finance while providing a clear overview.

Figure 6.3 **Project management tool.**

DIGITAL TOOLS AND FRAMEWORKS FOR DIGITAL TRANSFORMATION

Organizations are undertaking digital transformation initiatives across various business areas, including production management, sustainability, business insights and analytics, people and culture, reliability and maintenance, safety and risk management, etc. Customers across all industries are already on the path to digital transformation, and companies must follow. As it is an overwhelming transformation to perform, organizations are relying on different tools that support their digital initiatives. Besides the simple tools that can be purchased and modified to facilitate the overall efforts of different teams within the organization, many advanced emerging technologies might bring enormous benefits to larger organizations and those businesses that have a significant budget for their digital initiatives.

Digital twinning technology

Digital transformation is a never-ending journey. As new business models and technological solutions emerge, the journey will only be enriched. Artificial intelligence, blockchain, and the Internet of Things provide organizations with more value for their customers. Digital twinning technology is a convergence of several existing and evolving technologies, and it can be incredibly valuable in digital transformation.

Digital twins are created on the core concept of a digital form of something physical. From agriculture to automotive, every interaction an organization has with its customers involves physical entities. Digital twins are an exquisite opportunity for larger organizations to enjoy the benefits of serving better the needs of their digital twins. With efficient data management, organizations can form the foundation of any successful customer experience. Digital twins deploy the Internet of Things to collect real-time information from the world around us. This information is constantly processed, analyzed, and adapted to provide valuable insights. When an organization has access to real-time insights, it can also create and manage successful user-centric programs.

Also, digital twinning technology enables overcoming innovation barriers, mainly those related to high costs of failure. Due to a stimulating environment that combines real-time information, organizations can collaborate with the user community on developing quality offerings. This technology can also serve as a meta-layer that brings together numerous systems and processes in one place. Digital twins have a crucial role in streamlining business processes, from knowledge management to training and optimizing the business processes.

Data management software

Many small and midsized organizations run on old, well-known software suites, which no longer fit their current business needs, including working remotely. These software suites are often not well integrated, resulting in employees spending extra hours toggling between programs to support. Data management software should simplify the work processes of employees, not the opposite. Poor software management often results in poor relationships with customers, something all companies wish to avoid.

Numerous organizations are trying to update their software systems to avoid these problems, but they lack an understanding of building a digital organization from scratch. These organizations want to start with their digital transformation process but don't quite understand what it entails. After all, simply buying a new software suite will not make a significant difference if its tools are not integrated and work well with the existing tools and business models. Comprehensive systems integration and data management strategies are essential to building valuable customer experiences.

When purchasing data management software, an organization should focus on the centralized location where all data from all systems come in and are further processed. Another thing where organizations lose track with their digital initiatives is implementing the technology for themselves. Organizations implement each technology because of their customers, not themselves. Not to say that employees shouldn't see as many new benefits as customers, but the customer is the end goal. Many organizations implement a new technology that benefits their employees in a way, but customers either are not aware of this implementation or are having a hard time learning to use it.

Success in digital transformation implies platform selection, choosing the right set of basic apps, and the ability to quickly regenerate a running code to support all organizations dynamic processes. Another step that will lead to successful digital transformation is performing a needs analysis of each department and team an organization aims to transform. This is a very challenging task, yet it is necessary to ensure that everything implemented under the digital transformation umbrella truly serves the employees using these technologies.

Organizations of all sizes need to rediscover the importance of the human element in business and identify efficient ways of making the data management software support the most relevant business processes and interactions with customers. In the end, the ability of the software to function as a container for organizations can transform even the most traditional organizations by providing them with new insights on how to meet the needs of their customers.

REAL-TIME QUOTATION – DYNAMIC PRICING

What is digital pricing?

It sets prices for products and services using real-time digital tools and techniques. Using algorithms and automatization, prices are selected based on the current market conditions.

Figure 6.4 Digital pricing illustration.

Definition of digital pricing

Digital pricing is a process in which companies use digital tools and techniques to reshape their pricing strategies. This can range from building new pricing models to enforcing automated pricing systems. The ultimate goal of digital pricing is to improve business performance through more efficient and adequate pricing (Figure 6.4).

Digitizing pricing can offer numerous benefits, including increased accuracy, transparency, and speed. In addition, digital tools can help companies better understand customer needs and behaviors and enable more customized pricing.

Companies should consider digital pricing as it can help them tailor their pricing strategy to their specific goals and needs. Some digital pricing tips that all companies should consider are as follows:

- Use data analytics to inform your decisions: Gather data about your customers, products, and market conditions to make more informed pricing decisions.
- Be transparent: ensure your prices are straightforward and easy to understand, so customers know what they are paying for.
- Try different digital pricing strategies and find what works best for your business. Review your pricing regularly and adjust as needed.

How technology is changing the pricing process

Technology with A.I. is adjusting the way companies price their products and services. In the past, companies relied on manual processes for pricing. However, thanks to digital technologies, companies can automate pricing by leveraging data analytics and pricing models integrated with sales tools such as CRM and price quote generation (CPQ) software. This allows companies to approach pricing more strategically and better understand the impact of pricing on the bottom line.

Digitizing existing pricing processes

Today, digital technology is rapidly changing how businesses operate, and pricing processes are no exception. Many companies use digital solutions to streamline their pricing processes and increase efficiency.

Digital pricing solutions offer many advantages over traditional methods. They can help reduce costs, speed up decision-making, and improve accuracy. In addition, digital solutions provide greater flexibility and transparency, which can boost customer confidence.

Choosing the right digital pricing solution for your business requires understanding your needs and goals. Here are some factors you should consider:

Cost: How much are clients willing to spend on a pricing solution?
Ease of use: How easy to understand is the solution to use and implement?
Features: What features are essential to clients and their businesses?
Flexibility: How flexible does the solution need to be to meet your changing needs?
Support: What level of support does the merchant provide?

Digital pricing models

Companies can choose from several digital pricing models, each with its own advantages and disadvantages.

The most common digital pricing models are:

1. tiered
2. pay-per-use
3. subscription
4. freemium

Tiered pricing is standard among SaaS companies. Different features are offered at different prices depending on the level of service desired. Additional features are unlocked when the user pays more.

With usage-based pricing, customers are charged for each digital product or service they use.

With digital subscriptions, customers pay a monthly or annual fee to access a digital product or service.

With digital freemium pricing, customers can access a digital product or service for free but must pay for additional features or content.

A typical online fashion retailer has an assortment of 7 million items, while the big multi-category providers like Amazon have many more. Moreover, the most successful online retailers change their prices on individual items every 15 minutes. How is that possible? Suppose you're trying to figure out how many employees have to work tirelessly in their offices to analyze price elasticity trends and how they affect individual prices. In that case, you're barking up the wrong tree.

Dynamic pricing is fully automated, with computers calculating and processing vast amounts of data about competitors' prices, sales promotion figures, potential customers' search trends, product reviews on Internet forums, and even comments on Twitter and Facebook. Depending on the company's strategic goals, be it maximizing market share or profit, an algorithm calculates the optimal price on demand, sometimes updating every minute.

Dynamic pricing is critical in improving both consumer price perception and retailer profitability. Dynamic pricing, introduced by Amazon and others in 2005, is saving online retailers' profit margins as online shoppers have become savvy bargain hunters. Price comparison websites and review communities have enabled unprecedented transparency, and aggressive new retailers often blow up existing pricing structures with low introductory offers. Dynamic pricing can quickly generate a 3–8% return on sales and represents a tremendous competitive advantage.

Retailers counter with two different versions of dynamic pricing: one version optimizes prices for an entire assortment, which then applies to all customers. This is ideal for products that are easy to compare, such as branded items. The second version calculates an individual price for each customer. This works better when direct comparisons, such as for insurance products or travel.

Algorithms find the best prices

Fortunately for retailers, the best price is sometimes cheaper than competitors'. As local retail experiences also show, customers form opinions about whether a store offers good value for money or is expensive. Digitization is disrupting the traditional ground rules of retail pricing based on relatively few prices for specific products. In supermarkets, for example, prices on frequently purchased items such as milk, butter, and laundry detergent are usually marked up, so retailers select these products for their aggressive special offers.

And it's no different online. It comes down to knowing the products that customers base their opinions on value for money. The best example, once again, is Amazon: The e-commerce giant defines essential value items in each category that it consistently offers at lower prices than its major competitors. For example, in ink cartridges for printers, Amazon knows that most shoppers look first at the price of a double pack of black ink. In 2016,

Amazon undercut its two strongest competitors by more than 20% on this product. However, Amazon was slightly below its competitors on the single pack of black ink, which is also frequently purchased. However, these low prices are more than made up for by the prices of colored ink: yellow, blue, and red ink are between 33% and 57% more expensive than Amazon's competitors.

The great thing about digital pricing is that the company is constantly learning. Computer programs track customer and competitor reactions to a new price in real time: Are sales developing as planned? How many prospects did not buy? Where do the new customers come from – price comparison sites, competitors' websites, or did they come to our website specifically? All findings are immediately incorporated into the pricing model, which is constantly updated and adjusted.

Companies that optimize their prices according to this formula even do away with the old 80/20 rule, of which 20% of products account for 80% of sales and profits. Let us take the example of Amazon with its cell phones: About 80% of its revenue comes from selling devices, which account for about 20% of the business in that category. Accessories such as chargers, connector cables, headphones, and smartphone cases account for the remaining 80%. Although they account for only 20% of sales, they contribute 50% of profits. These are typical long-tail products. They remain in the range for years, even if the corresponding cell phone has long since been superseded by newer models but is still used. Stocking such an extensive range of accessories in stationary retail is not profitable because the storage space is too expensive. Online retailers, however, with their vast and inexpensive warehouses, can healthily increase their profit margins with just these items.

All major online retailers have now introduced dynamic pricing systems, and the idea is also catching on in brick-and-mortar retail. For example, U.S. retail chains Sears and Home Depot can directly change their in-store prices at the touch of a button after installing electronic price tags in some departments of their stores. Once electronic pricing becomes widespread, opportunities for dynamic pricing will also open up for multichannel retailers that started in the physical space. After all, today's customers expect a consistent offering across all channels, including prices.

Customized pricing

Customized pricing goes one step further. Here, the retailer tries to classify individual customers by analyzing, for example, which device they use to access the website. If an expensive iPad is used, the system will immediately display a higher price than for another customer using a cheaper product with an Android operating system. A more reasonable offer is also displayed to customers who were redirected to the vendor's website from a price comparison site. For a long time, systems had remembered whether someone had visited the website before and was interested in an offer. If the customer looks at the same item on the website again, the system forces a decision by promising a discount for immediate purchase or offering free additional services.

When customers see through these pricing strategies, they often need to be more enthusiastic, which is why many travel companies have given up on differentiating customers by the devices they use. Too many customers have expressed their frustration at being shown a higher price after glancing at their expensive iPhone than they were later shown at home on their old PC. However, price differentiation by route to the website is still very much in place among the leading players in the travel industry.

The concept of customized pricing is of interest to companies in many industries. Insurers are working on better pricing individual risks, while energy companies are looking to incorporate personal consumption habits into their offerings. And it's not just in retail that dynamic pricing is catching on; companies are also experimenting with the concept, for example, in the chemical and steel industries. Steel trader Baosteel's online marketplace Ouyeel has created price transparency not previously seen in the industry. And BASF now trades on the Chinese trading platform Alibaba, selling chemicals to thousands of primarily midsized customers in Asia.

Five modules of dynamic pricing

Dynamic pricing is critical in improving both consumer price perception and retailer profitability. A robust dynamic pricing solution should consist of five modules, all working in parallel to generate price recommendations for each SKU in the assortment:

1. The Long-Tail module helps retailers set the introductory price for new or long-tail items through intelligent product matching. The module determines which data-rich products are comparable to new things (with no history) or long-tail items (with limited historical data).
2. The Elasticity module uses Big Data analytics and time-series methods to calculate how a product's price affects demand, taking into account various factors.
3. The Key Value Items (KVIs) module estimates how each product affects consumers' price perceptions using actual market data rather than consumer surveys. This allows the module to identify which items consumers perceive as KVIs automatically.
4. The competitive module recommends price adjustments based on competitors' prices updated in real time.
5. The omnichannel module coordinates prices between the retailer's offline and online channels.

Although a best-in-class solution includes all five modules, companies can often start with just the KVI and Competitive Response modules. These help companies respond quickly to competitive changes on key items while adding the other modules.

Companies that want to lay the groundwork for a functioning dynamic pricing system know that it can be done – and that the effort is worth it. Some e-commerce retailers have increased their margins by two to three percentage points – in an industry where margins are tight. This can make the difference between an industry leader and a follower.

And with results like that, concepts catch on quickly. Over the next few years, dynamic pricing will likely become a core business competency. The next wave is likely to affect the B2B sector, where pricing still needs to be transparent in many industries. This new openness could send shock waves. Just think of the current practices in industries such as steel or chemicals. In the B2C sector, the trend is increasingly toward customized offers, including tailored prices.

The big question is whether customers will rebel. Just because something is technically feasible does not mean it will be accepted, as the example from the travel industry shows, where companies have had to abandon price differentiation depending on whether a customer uses an expensive Apple product or a cheap no-name device PC. But there are other ways of setting prices that will pose an exciting challenge for creative companies.

IN THE HEART OF THE COMMUNICATIONS ECOSYSTEMS

The trigger for the revolution is tiny: an area of just 6 by 5 millimeters and 1 millimeter thick. These are the dimensions of the eSIM. Like its larger predecessors, this new memory card ultimately establishes the connection between mobile devices, the Internet, and the mobile network. What is revolutionary, however, is not so much the tiny dimensions but rather the "e", which stands for "embedded".

Currently, Subscriber Identity Module (SIM) cards are shipped by a mobile operator and usually inserted manually by the customer. The programmable eSIM is installed by the manufacturer in smartphones, tablets, fitness trackers, smartwatches, gaming consoles, smart glasses, cameras, or home medical devices – constantly online. With the eSIM, users can also make phone calls via wearable technology. The classic 3G smartwatch, Samsung Gear S2, for example, has an eSIM that allows users to choose their mobile carrier, make calls, and access the Internet independently of a phone connection.

The prospect of this excites customers but worries telecommunications companies, which played a dominant role in the early days of cell phones when transmission capacity was low, and they were in control. The telecom companies sent their customers SIM cards and were in a central position with high added value. If the cards from SIM can be assigned to any provider in the future, the mobile operators will lose their most important lever for customer retention – the cost of switching. This will lead to a redistribution of profits and revenues in the relationships between telecommunications groups, hardware manufacturers, Internet providers, and content providers.

Battle for the new ecosystems

New ecosystems are emerging, and all market players want to be at the center, where value creation is most significant, rather than at the periphery, where they feed off the scraps. Will hardware manufacturers like Apple and Samsung win the day? With eSIM, these companies can pre-determine the wireless carriers for their smartphones and allow customers to switch carriers with a simple click.

It is also a logical step that the hardware manufacturers themselves could buy transmission capacities and infrastructures on the market and displace the wireless providers. Or will "content is king", a philosophy espoused by content providers like Netflix, prevail? The video streaming service relies heavily on self-produced series and movie TV and has attracted millions of new customers. As a newcomer to the market, Amazon is now also producing its content. Facebook is also betting on virtual reality (VR). The social media company bought Oculus, the maker of virtual reality headsets, for $2 billion. The Oculus Rift offers all kinds of digital experiences, from virtual car chases to full virtual tours of properties for sale. The device can trick the brain into thinking that the body is experiencing what we see (Figure 6.5).

Several players have entered this market, the most successful of which is Sony with its PlayStation. Together with the related field of augmented reality, an attractive growth market has emerged. According to a study by Goldman Sachs, the new industry, which already generates between $3 billion and $5 billion a year, will grow to between $80 billion and $100 billion in revenue by 2025.

Although gaming accounts for nearly all revenue, business users are expected to account for about half by 2025. VR Headsets can help designers by displaying virtual information, help surgeons by showing virtual lines along which they move their scalpels and even help soldiers aim their rifles.

The telecommunications giants have a lot to lose. Their once-lucrative calling, messaging, and video services have long been offered for free by aggressive competitors. They are also on the defensive when it comes to content and platforms. Since 2013, the total revenues of telecommunications companies in the U.S. and Europe have been falling by around 0.5% a year. And this decline is now threatening to escalate. Various worst-case forecasts predict a decline in industry revenues of up to 50% by 2025.

To stem this trend, some companies are using their still-healthy cash resources to buy into companies that produce content and programming that can be distributed over the telecommunications infrastructure. In late 2016, U.S. telecom giant AT &T made a $85 billion offer for Time Warner, whose portfolio includes movie studios and the CNN television network.

Figure 6.5 SaaS players.

Earlier, AT &T had just completed its acquisition of DirecTV for $49 billion. A few years earlier, cable provider, Comcast acquired media group NBC Universal. And in 2015, wireless provider Verizon acquired Internet pioneer America Online (AOL), and later, in 2016, Yahoo!

More than acquisitions alone, however, will be needed to save the industry's profitability. To ensure that telecoms do not end up as low-paid commodity suppliers of basic infrastructure, they need to improve in three areas:

1. They need to streamline a core business that has become slow and cumbersome over the years.
2. They need to identify growth markets and develop strategies to capture them.
3. They need to get a handle on regulation, as industry structure and consolidation will be critical factors.

Streamline core business

Reducing costs can be helped by digitization – the phenomenon that has caused the need for leaner processes. From customer acquisition, registration, payment processes, billing, and customer support to contract cancelation, every step of the Customer Journey is now under the microscope. Each step of the Customer Journey offers the opportunity to replace expensive human labor with digital assistants, with the goal of digitizing the customer contact process end-to-end, with a higher level of service and lower costs than before.

Of course, management and technology must also be streamlined. Surprisingly, these new digital processes are more cost-effective and usually lead to higher customer satisfaction when implemented correctly. Instead of being put on hold by a call center, many customers are happy to resolve their questions digitally.

Opportunities in new service sectors

After streamlining their processes, telecommunications groups can tackle future growth markets. Wireless providers are in an excellent position for six key growth markets because their networks are ready for the data flows.
These markets are as follows:

- **Wearables:** Who will connect fitness bands, VR headsets, smartwatches, and running shoes to the Internet? How will their revenue be generated, and what is the business model?
- **Smart homes:** Which Internet connection will control the heating and air-conditioning systems, shutters, elevators, and all the other functions of tomorrow's smart homes?
- **Connected cars:** Autonomous driving, lane departure warning systems, emergency braking systems, service data – who will process the enormous data stream that the vehicles of tomorrow will produce?
- **Internet of Things:** All machines are equipped with sensors that constantly transmit performance and usage data – who will ensure the transmission of this vast pool of data?
- **Digital health:** the networked patient will soon be transmitting a constant stream of data – will this be done via the networks of telecommunications companies?
- **Cloud computing:** Who will run the data clouds – the data and software centers where all the data will be transmitted and processed?

The key questions are as follows: What role will telecommunications companies play in these growth areas? Will they, as in the past, only supply the transmission technology that will be cheap and interchangeable in the future? Or will they succeed in assuming a central role in the emerging ecosystems?

Will they be able to put their SIM cards into the devices, or will they be displaced by the manufacturers who will then have control over the mobile provider? Will they pass the data to other cloud operators' data centers, or will they be able to provide the necessary data centers and software services? And will they be able to build up expertise in data analytics and offer services to their external partners in the following areas?

AT &T, for example, is eagerly tapping into new service areas and has established itself as the hub of a new ecosystem. The company is leveraging its large customer base and the trust it has built over 100 years to offer security and convenience packages around the home. For subscriptions ranging from $30 to $65 per month, residents' windows and doors are monitored when they are away. A local security company is notified and checks the premises to see if a window or door is opened. An alarm is also triggered by smoke, fire, or flooding. In terms of convenience, the packages also include remote control of heating and lighting and monitoring of pets and babysitters via camera. Customers can see the camera on their smartphones or tablets while away and also control the functions of the various systems.

To make all this possible, sensors, cameras, and actuators must be installed in the home by contract providers from AT &T. Along with the subscription fee, this gives the company one-time payments of $30 to $150 while making it clear to the customer that the equipment is worth much more. AT & T manages its network of equipment partners, local installers, and local security companies, and – as the provider of the offering and controller of the customer relationship – can claim the lion's share of value. And in an era of minimal fixed-line telephony revenues, predictable annual revenues of $400 to $800 per customer are expected. For this reason, the idea is already being copied: Swiss telecommunications company Swisscom has a similar offering called Smartlife.

The other players in this emerging ecosystem are also doing the math. Google and Apple are already equipping their mobile devices with an eSIM and promising their customers a seamless switch from one mobile provider to another. They are also pushing into new business areas. Google tried to get a piece of the smart home pie in 2014 when it bought Nest, a connected home company that had already established itself in the market with its smart thermostats and smoke detectors. Traditional players in this market, manufacturers of everything from heating systems and thermostats to lawnmowers, are also looking to gain a foothold in these new ecosystems and connect their devices.

Chapter 7

Case study

Digital transformation in logistics company

COMPANY PROFILE

Lighting company which has a strong impact in Croatian market as distributor for major brands. Doing business more than 25 years mostly on local market but raising numbers in export to foreign EU countries. Currently there are 15 employees within two branches in bigger cities. Average income in last three years was over 3.000.000 €. Company doesn't have strict structure apart from CEO/owner as the main responsible person. Most of the employees are educated faculty, having a long experience in logistics field of business.

CHALLENGE AND DIGITALIZATION STATUS

For business purposes, company has five cars and three delivery vehicles. Client/partner visits are handled by monthly schedule to gain sales and trust. Other orders are realized by delivery services towards mainly other business subjects (some small revenue comes from private persons). The firm has been moderately digitized for the past ten years. Along with relatively newer computer equipment/server infrastructure, it owns an up-to-date website. Recent interventions in terms of digitization went in the direction of unified e-mail addresses (common address book) and tracking of received and sent e-mails (common e-mail boxes). Every employee has company mobile phone and laptop which can be taken on a business trip or home office (Figure 7.1).

Since basis for digitalization was set, owner decided to make the next digital transformation as a mini step. The mentioned vehicles were in focus where expenses weren't tracked or analyzed regularly. Every vehicle cost/expense was defined as follows: vehicle purchase, all maintenance (regular, unplanned, preventive), fuel, cleaning services, checking, or monitoring the pressure in the vehicle tires as well as windshield washer fluid refill, fire

DOI: 10.1201/9781003305163-7

Figure 7.1 Digital transformation in logistics.

Table 7.1 Data challenges among employees before digitalization

Challenge	Description
Lack of information	Vehicle cost analysis is challenging since it would cause separate tracking in posting process. Quick cost and effectiveness picture was unable to get. Idea was to generate personal vehicle ID where each one will be having crucial data, handling record, as well as current value in comparison to invested.
Lack of employee attention	Since everyday work tasks are overwhelming, it was easy to forget deadlines for vehicle maintenance (ex. Service due, fire extinguisher check-up etc.). Digital solution would cover all mentioned situations together with other repetitive tasks which employees could easily forget.
Vehicle indebtedness	All vehicles were used by different employees (no specific vehicle was appointed to only one employee), so lack of responsibility occurred in case of accident or vehicle misusage. Application had an aim to appoint each vehicle to employee in hours/days when he/she is using it.
Reports	In case of any damage, insurance companies ask for photo documentation. Supervision by the management is also easier if every defect is properly noted when using or taking over the vehicle (ex. the previous employee left the vehicle in an untidy state the day before).

extinguisher maintenance, radio/service subscriptions, all potential crashes and vehicle damages. Notation for all those facts would provide better insight into company assets together with efficiency assessment. Situation before digitalization is noted in Table 7.1.

SOLUTION – APPROACH AND REALIZATION

As one of most popular ways in digitalization approach, owner of this company decided to give contract to a software company. It won't be appointed just to produce a solution but to help in better understanding and long-term development in this new, refreshed process. Owner and his present team described problems which was the base for technical documentation for this project. According to the suggested plan there were phases prepared with defined outcomes. Whole digitalization process was estimated to be finished within eight months.

After preparing detailed technical specification, the software company tried to offer a faster solution by researching similar commercial and open-source solutions. It was concluded that no application can cover specific needs of the company. Some applications were too wide and others too narrow in functionality terms, so it was decided to go with custom development. Such approach is more expensive since development is devoted only to one client, but as such it solves all predicted challenges for a specific company. Web and mobile applications were the expected outcomes of the IT project.

For the technical platform, Laravel environment was accepted due to its wide usage, standardized code libraries, and ability for owner not to be dependent on only one vendor (software company). The agreement pointed that provided code after solution is implemented will be in possession of the client where owner would be able to decide to change software company if needed. MySQL database was chosen for data storage where it was seen as optimal and easy scaled solution for this project. Cordova framework was chosen for mobile application which had considerable saving (in time and cost) in comparison to native development (so separate for Android and for IOS operating systems).

This specific case was straightforward to explain and understand for developers. During design and development phases, both sides (client and software company) were involved especially in coordination about user experience which needed to be as simple as possible. After first development iteration (within six months) and first alpha version build, a couple of company employees were selected to do an extensive test. Feedback confirmed that current progress is satisfying, and that process is moving into right direction (Figure 7.2).

Additional two months of work was invested in order to provide first stabile beta version. Testing period started where developers were involved and physically participated with the client in real-life scenarios suggesting an optimal approach in note taking in application. This case assumed involvement of almost all employees in testing period, so education was relatively simple.

Doing such approach also minimized resistance to change since all employees were aware what is going on, why such application is created, and what benefits it will provide. After successful testing time frame (two weeks),

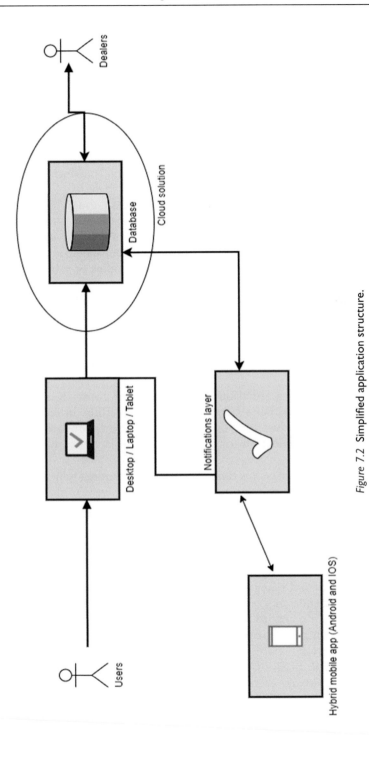

Figure 7.2 Simplified application structure.

it evolved to production usage meaning that all company members have started to use an application, are getting used to it, but also taking notes of what can be additionally improved.

Last mentioned thing is important since it provides a feeling of an opportunity to whole staff to actively provide solutions for their everyday challenges. The company management decided to award best improvement ideas every month so that it becomes an extra trigger for employee engagement.

Application benefits which company management planned were achieved. Usage within six months provided saving (vehicle service prediction and repairing analysis which showed vehicle unprofitability) which led to almost half of the project investment return.

During couple of months of active application usage, all employees accepted it and together with the company management started to plan for further development. Together with partner companies, update which would cover working machinery, trucks, and other construction machinery were prepared. Covering support for motor bikes or e-bikes was also considered.

Extra point which is recognized as needed during implementation is an option to connect with ERP systems. It would provide an option to automatize cost invoices towards accountant process (ex. If an employee pays for the gas, by simple invoice scanning it would become available as dealer invoice in accounting module – no matter if the cost is paid by the employee or the company).

This case showed a situation when no current commercial product is suitable for company digitalization process. Almost like in real life, if you are tailoring a suite just for you, it will fit the best. Similarly, if you are building custom application for your firm, it will have an optimum set of needed options.

However, such approach is time and money consuming where it is easy to under/overestimate it in every resource you may have. On the other hand, experience showed that by doing so, companies which can afford it are better prepared for further digitalization and full digital transformation steps. By incremental development of digital activities within different departments, it is easier to adopt new ways of business.

It is also extremely important to emphasize employee awareness and preparedness for digital renewal phases. Often, company management tries to do so by using shortcuts or as a must. A lot of resources in such approach are wasted when trying to minimize resistance. Without clear vision about digital transformation propagated within all company members, it could fail or last forever.

DISCLAIMER: In all case studies, company names were removed due to privacy and company secret issues. Examples are provided from authors' experience and are used only as proof of concept what was happening during digitalization in different companies and their business processes.

Chapter 8

Case study

Digital transformation in manufacturing company – paper products

COMPANY PROFILE

Manufacturing company as Croatian leader in industrial paper bags. Firm and management have extensive experience not only in local but also in foreign market. Since 2005, the company is constantly growing with current number of 50 employees. Average income in the last three years was over 5.000.000 €. Company has five leading roles in management with additional three responsible persons within departments (mostly warehouse management). Most employees have a high school degree, but extensive informal education in machine handling and paper production (Figure 8.1).

CHALLENGE AND DIGITALIZATION STATUS

As the company was growing, new production technology was adopted together with information and communication solutions. Management and department leading roles are having most data-driven positions since every paper bag produced counts in terms of efficiency (both, machine, and employees).

IT equipment is relatively new, consisted of laptop for each mentioned employee together with external display. Such approach has proven itself as optimal since it combines two principles (laptop and PC). In case employee needs to go on a trip, laptop can be easily carried with him/her having all data needed.

On the other hand, working in the office is not exhausting for the eyes since large display is on the desk. Easy "plug and play" approach with IT dock solution (device which connects all periphery as keyboard, mouse or printer, LAN...) can be attached with one click (USB-C or connector below the laptop). In every three-four-year span, management is investing an effort to renew the equipment considering all current trends.

Since company location is in suburban area of the city, it lacks broadband Internet connection. Maximum solution was 5 Mbits/sec download and 1 Mbits/sec upload which is way too bad and far from optimal.

DOI: 10.1201/9781003305163-8

Figure 8.1 Manufacturing and digital transformation.

Before any serious digitalization process company tried to find a better solution. Internet Service Providers (ISP) were not prepared to find fully stabile solution via infrastructure cables like fiber-optic but managed to provide (temporarily) 4G solution where 50/50 Mbits/sec was achieved. Considering all employees who are to be connected constantly to Internet (approximately ten), this solution is as good as it gets.

If we divide a speed with employee number, we'll have 5/5 Mbits/sec. However, some company workers will be connected over VPN which needs more bandwidth speed together with main server. Since 4G, as mobile network, depends on usage, location, and other parameters, it is expected that in some periods of the day speed will be slower. As a temporary solution, this was only possible approach with strong guidelines to ISP to provide more stabile connection as soon as possible.

Company owns modern web site which is not updated too often. Main communication channel apart from mobile phones is the email where every employee has unified address name.surename@company.com. As opposite, some companies are choosing name@company.com or nick@company.com which is considered as more flexible and relaxed approach.

Also, the company has own local server which handles accounting software and is used as data storage. By using a domain controller, each employee once logged on to the computer has set of rules and permissions what can be used. Server and each employee data are securely backed up on three different locations. Two are off site on distant server and one copy remains on additional external NAS.

Main data which almost the whole business relies on consists in Excel sheets and that is the main starting point where new cycle of digitalization has started. We can consider Excel sheets to be the first stage of digitalization process since those could contain crucial data which is to be processed in further stages. On the opposite side, it could lead to a dead end since business owners are used to such tables and sometimes aren't able to see the "wider picture" once the new solution is on the horizon.

As described in Table 8.1 (and that's only a part of full data which was provided in sheets), client had a relatively good start when it comes to a specification what needs to be reported from data. The downside of this was that data was manually entered with large space for error. Main idea was that automation will provide data in application which would ensure raw input for calculations.

Another challenge appeared once machinery was examined. Since some of the machines were dated three decades ago there was no interface which application would connect to. The only option was to build a custom solution for that as well. After additional research it became clear that input digital boards should be purchased where the relay will be connected. Outdated technology with a little bit of electronics was updated and the engaged development company managed to gain the results into the database (Figure 8.2).

In similar way client has solved connection to a semaphore with lighted errors which machine produced. New data allowed tracking of all errors (number of occurring, error lasting). Preparation of exact data was the first project phase. After that, Excel calculations were a scenario for technical specification. Since all data was gathered on local area network and databases on local computers, it made sense to make an internal interface. Additionally, the board was able to approach the interface via simple browser by adjusting VPN access into company. Directors who were away from the factory were able to see real-time results.

The web application and its interface made an extra step in efficiency tracking. Not only that raw numbers were reported, but a comparison among workers, shifts, or working orders were prepared. Company board decided to use those metrics in salary calculations. Huge TV placed in factory discovered speed and current shift efficiency so positive competition

Table 8.1 Data consisted in Excel sheet

Data	Type	Note
Bag production	Number	All data is structured in Excel columns. However, human errors on input are causing wrong or uncertain calculations as well as reports. The company board would like to use this table as a base for a salary.
Workers	Names	
Client	Text	
Work order	Text	
Total time invested	Time	
Time of errors	Time	
Worker efficiency	Total bag production/Invested time + Time of errors	

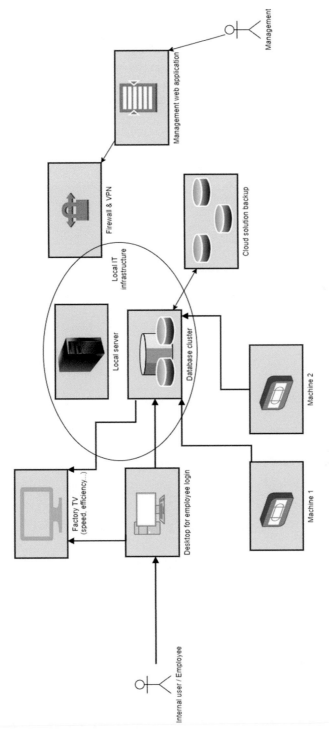

Figure 8.2 Application structure showcase.

among employees occurred where bigger number meant being extra careful in machine preparation but also a bonus from the management.

Even though in this case company has just started own digital transformation path by doing it a proper way, results started to grow in just couple of months. Definitely, there's more room to develop not only this efficiency system, but now employees are eager to suggest own solutions for their everyday tasks.

Chapter 9

Case study
Digital transformation in food company

A large worldwide corporation decided to cut off one part of food production since it was not lucrative enough. That same business meant a lot for another foreign company, so it decided to take over the job, clients, and know-how. New company realized that moment as a great opportunity to make a deep digital transformation.

From the beginning, it was clear that migration from one corporation to another won't be a smooth one. However, managements got together and tried to do a timetable and schedule on this process. Main "black box" for a new company was what's current digitalization status of production processes. Two additional IT teams were hired from four different countries which will supervise this process as well as be "eyes and feet" for engineers in new corporation headquarters. One team was leading infrastructure tasks and the second one, software solutions (Figure 9.1).

Since main product was an ice cream, apart from technical segment, there was a technological process which was there to oversee and check in detail. Food engineers had a special training there where they needed to get familiar with all procedures and guidelines from the old company.

Once food is in focus no compromises in terms of quality and rating can occur. Similar is with IT in that department. Ice cream during its production and especially in phase of storage must obtain in low temperatures. Electric equipment and low temperatures aren't a good fit. New corporation wanted an IT-enhanced overview of each produced ice cream box. That way, better understanding as well as value of stored good will be available. Solution was to go with QR codes which contained a lot of product details together with tracking details, ingredients, and date produced. Second IT team worked hard to provide all those information and to assure data consistency and accuracy.

After more than a week of testing, whole storage labeling system went down. Main router in the biggest warehouse got frozen since inside temperature was −10 degrees Celsius. Just to analyze this issue, teams had challenges since outside temperature at that moment was approximately 35 degrees Celsius. Going a couple of times, a day within a range of 40 degrees was a recipe for catching a cold.

DOI: 10.1201/9781003305163-9

Figure 9.1 Digital transformation in food company.

In the end, team decided to go with custom-build router case which would allow it to be nice and cosy no matter on outside and inside temperatures. Predicted Wi-Fi signal shortage occurred because of the case, but the team was prepared with two extra Wi-Fi extenders which needed to ensure network coverage. Forklifts appointed to distribute large ice cream box to transport ramps needed to be connected to Internet no matter on exact location. Extenders did the trick, so with extra purchase router on each warehouse team was sure that there will be no networking problems no matter on weather or forklift distance.

Another challenge for second IT team was software. Since new corporation had own standards and previous good practice with some IT solutions, idea was to implement them in this business as well. However, relatively old, and unmaintained equipment – servers in the first place, allowed no room for updating or expecting it to work in the long run.

Detailed equipment evidence became very important. Since previous company had some reports with questionable rate, new corporation decided to invest substantial amount of time with IT team to clarify what's the equipment condition. Desktops, laptops, network equipment, and even tablets together with IP phones or printers were on that list, but no one knew where exact on the site. After two weeks of exhausting work, corporation had a clear overview that just a couple of devices for middle management should be bought and everything else is enough for mid-term digital transformation (2–3 years).

All equipment was labeled with standard barcode. Corporation decided to add all noted things within asset management tool used on all company locations – Snipe IT (https://snipeitapp.com/). Such solution was used first time two years ago and since then considerable amount of time was saved in finding spare parts, knowing which device is outdated or too old to repair, etc.

Almost all corporations are using local vendors once they talk about local legislative in the country, they are doing business in. It would be crazy to have a dozen solutions for every country multinational company operates in. Finding a right vendor is also not a breeze, especially knowing that those guys will be responsible for salary check or any other labor law issues which might arise. Smaller companies either love or hate working with large corporations. Final decision is up to their strategy and vision of future work.

For a corporation is always a good thing to have flexible and agile team which would answer to their needs (in anytime). Financial aspect and credibility of large corporation for a small company is definitely a benefit. However, it can be always challenging to depend on team consisting of two–three people, especially if they built custom solution which no one can maintain (Figure 9.2).

In the end, large corporation decided to go with medium-sized company as provider of local IT system dedicated to handle HR, payrolls, time tracking, and simple accounting which they didn't have in large ERP system. Hired software consultants were engaged in eligibility check where one of the main issues was translation of local IT system which wasn't fully finished. Since local company was eager to provide their solution to such large corporation, they finished translations within two weeks. After showcase of all needed explanations, they gained long-term clients. New modules as well as feature updates will be implemented within maintenance agreement where fixed number of working hours is included. Every extra hour is billed separately after appointed person's confirmation.

Large companies can contain huge issues during transformation process. Since in this case, abroad corporation had some previous experience in it, good preparation was made so a lot of challenges were minimized that way. Quality team of consultants assured that infrastructure, networking, desktop/laptop equipment, and software solutions are on scalable (as much as possible) and acceptable level. It is always interesting to see how large corporation handles outsourced companies. Especially when the deadline is short like in this takeover process. Apart from local IT system vendor, corporation searched for IT technician as employee. Since a lot of requirements were set in application, no suitable candidate was found. After further decision-making, another local DevOps company was found with ten experts which were and still are able to provide high-level and timely service for all corporation needs.

Since company is successful, constant adaptations are always implemented. Current main goals in next two years are better connection with regional market with stronger CRM solution with social media coverage. On the other hand, scaling infrastructure in terms of new servers is planned. Experiments with artificial intelligence are in progress where corporation is seeking long-term partners which will provide more optimizations regarding on big data analysis they can provide.

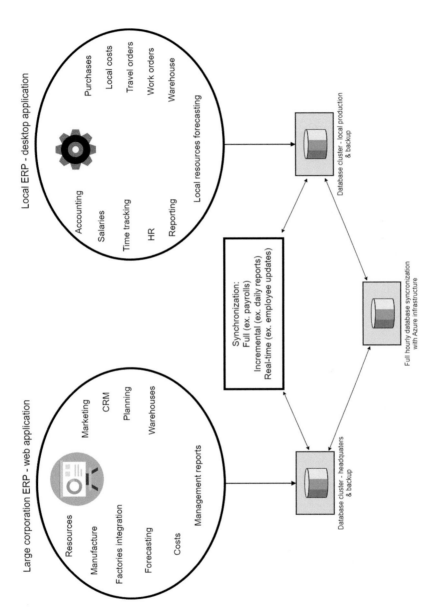

Figure 9.2 ERP comparison between corporation and local version.

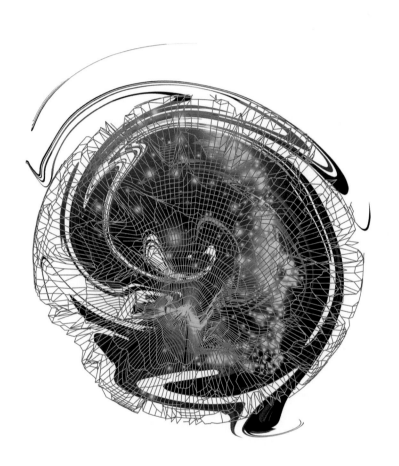

Chapter 10

Case study

Digital transformation in healthcare industry

Digital transformation impacts almost all industries, and healthcare is no exception. Now, it is more crucial than ever that the industry offers better patient care digitally, wherever the patient may be. In general, the adoption of cutting-edge technologies for enhancing hospital worker productivity and elevating patient care to a new level is where digital transformation in the healthcare business brings the most value.

By integrating and putting into practice cutting-edge technologies, digital transformation in healthcare offers a new method to provide care while improving internal hospital procedures and, most importantly, addressing all patient needs. It is the intentional application of technology to improve the lives of individuals and the efficiency of healthcare institutions (Figure 10.1).

The main objective of digital transformation in the healthcare industry is to implement revolutionary healthcare IT solutions to develop the most efficient workflow processes and compliances and enhance patient service while decreasing costs. The COVID-19 pandemic accelerated the number of healthcare trends, specifically consumers increasingly focusing on convenience and access to care.

In functional areas, several hospitals and health systems have incorporated digital technologies over the past 20 years. While primarily concentrating on the same business and customer models, they frequently adopted a piecemeal approach to multiple efforts, from installing electronic health record (EHR) systems to developing applications to experimenting with disruptive technologies like artificial intelligence (AI).

Leading healthcare industries view digital transformation as a way to become more consumer-friendly while simultaneously changing their operations, culture, and use of technology. Just by quickly looking at the industry, one can notice many opportunities in the diagnostics and healthcare sectors related to digital transformation.

For the healthcare sector, the COVID-19 pandemic drastically changed this state of affairs. Virtual healthcare and treatment provided in the patient's home became the standard for both necessity and preference. However, this transition was not as abrupt as it may have appeared. So much has changed due to the COVID-19 pandemic, such as changing consumer tastes, quickly

 DOI: 10.1201/9781003305163-10

Figure 10.1 Digital transformation in healthcare.

developing technologies, emerging talent models, and therapeutic innovation. Hospitals and health systems will likely focus their transformation strategy on a well-defined approach to digital technology as they try to change their businesses in light of these trends.

HEALTH INFORMATION SYSTEM

A health information system (HIS) consists of components and procedures created to generate data that may be utilized to enhance decision-making throughout the healthcare industry.

It provides the underpinnings for decision-making and has four crucial functions: data creation, compilation, analysis and synthesis, and communication and usage, based on the most commonly used definition by the World Health Organization.

Systems that manage data relevant to the operations of providers and healthcare organizations are included in the category of HISs. These could be utilized for research, enhancing patient outcomes, policy, and decision-making. Security is a top priority since HISs frequently access, handle, or keep significant volumes of sensitive data.

Data are transformed into information for health-related decision-making by HISs, which gather data from the health and other relevant sectors, analyze it, and verify its overall quality, relevance, and timeliness. Understanding how data systems are utilized, as well as user and manager behaviors that may affect performance, is crucial for improving HIS since it provides key information on how to adapt it. In this equation, people are at the center.

It is impossible to simply introduce technology without understanding who is responsible for running health systems and who can encourage change.

These aspects are also reflected in data collection and analysis. The engaged employees must be motivated and aware of the necessity for developing processes.

Investing in digital health

Global connectivity and the expansion of information and communications technology can accelerate human progress, close the digital divide, and create knowledge societies. Despite advancements, many nations still require institutional support creation and fusion of national e-health and digital health strategies, as well as the execution of their action plans, which typically calls for more resources and capabilities.

It is crucial to engage in and coordinate efforts locally, including those of governments, donors, and the corporate sector, with a broader approach rather than a narrow concentration on technology to sustain advances.

It is time that digital health is integrated into health priorities and serves people in an ethical, secure, reliable, equitable, and sustainable way. The pandemic revealed many pre-existing vulnerabilities in health systems and the importance of having high-quality data available for decision-making.

DIGITAL TRANSFORMATION TRENDS IN HEALTHCARE

Utilizing cutting-edge technology, techniques, and processes, digital transformation in healthcare aims to provide patients, healthcare providers, and healthcare organizations with lasting value.

Digital transformation involves accepting and using cutting-edge technologies in creative ways to optimize their benefits. A recent survey by Deloitte found that 92% of healthcare institutions and professionals improved performance because of digital transformation. So, what are the implemented healthcare trends that resulted in success? Which trends are reshaping the healthcare industry, and how?

Automation

Automation is the most effective tool available to the healthcare sector. It is important to note that its adoption is quite advantageous for drug producers, particularly concerning product safety. Technologies make it possible to examine materials with great depth and accuracy. Microbiological studies can benefit from using automation algorithms because they can increase study productivity, test and evaluate specimens, and enhance the quality of laboratory research.

There are numerous benefits to healthcare services. Automation helps workers reduce paperwork, lower human error, and increase overall departmental production.

For doctors, it entails having more free time at work, allowing them to interact more with patients and deliver even better, more individualized treatment. Additionally, automation frees up time for an individualized approach and allows caring for larger groups of patients with comparable needs. Additionally, such patients can get automatic appointment reminders, which reduces hospital no-shows.

The following two long-term requirements make digital transformation and healthcare automation necessary:

Increased patient expectation

The demands that patients have of the healthcare industry have drastically changed. Patients now demand more convenient and individualized care. Not only that, but most of them increasingly favor digital healthcare over offline appointments, particularly as a result of the COVID-19 pandemic.

Increased chronic diseases

The strain on healthcare organizations is rising due to the development of chronic diseases. For instance, estimations say that 415 million people worldwide have diabetes. By 2040, this figure is expected to reach 500 million, according to the Centers for Disease Control and Prevention.

As a result, automation and digitization are essential to maintaining a viable healthcare institution since they can aid in disease detection or shield people from chronic illnesses.

Healthcare automation is accelerating digital transformation and bringing forth significant innovation and advancement in the sector. With the help of numerous automation techniques, healthcare facilities may dramatically enhance employee and patient satisfaction.

Automating administrative activities can help operating processes run more smoothly. Digitization, for instance, can assist teams in streamlining paperwork, speeding up patient physical examinations and enabling more simple and convenient access to precise patient health data. By gathering and examining data obtained using digital tools, healthcare organizations can comprehend the demands and behaviors of patients. That may lead to developing fresher ideas for implementing value-based treatment and fostering patient confidence.

Automation leads to more efficient communication. For instance, online tools make it simple for doctors to communicate with their patients. Also, it is possible to improve internal communication inside the organization. Medical practitioners may communicate with one another more quickly, make decisions more quickly, give better patient care, and produce better clinical results with online platforms.

Yet, organizations are not the only ones who can benefit from automation. Using digital platforms, patients can post questions anytime and get immediate answers from their doctors. Additionally, automation offers better, more individualized healthcare services, allowing for more precise diagnosis and treatment for patients. It also makes access to private health records simple. In the convenience of their homes, offices, etc., patients can monitor their health metrics and data by utilizing their smartphones or PCs.

Connected ambulance

While driving the patient to the required department, a connected ambulance assists healthcare providers by gathering and sending all essential patient data gathered through wearables, sensors, and HD cameras to the hospital.

Doctors will have all the information they need to perform the necessary operations faster and more successfully without wasting valuable time even before the patient arrives at the hospital. In some specific situations, doctors can instruct paramedics on ways to perform specific tasks using linked ambulance technology.

Care must be delivered outside of hospitals to raise the standard of treatment and lower death rates. Improving the capabilities of ambulances through connection seems to be a promising way to do that. Healthcare services might be significantly enhanced by an ambulance fitted with telemedical tools that enable the ambulance crew to diagnose or treat a patient efficiently on the spot.

One of India's leading communication solutions providers, Bharti Airtel, collaborated with Apollo Hospitals, Cisco, and others to develop a 5G-connected ambulance that will revolutionize access to healthcare and save lives in dire situations.

Modern medical technology, patient monitoring software, and telemetry devices are all included in the 5G-Connected Ambulance. This technology can transfer patient health data to the hospital in no time. Additionally, it has camera-based headgear, bodycams for paramedic staff, and onboard cameras, all connected to the lightning-fast Airtel 5G network. According to the company, the plan is to enhance it using technologies like AR and VR.

When a patient in critical condition is en route to the hospital, every second matters. The 5G-Connected ambulance is an extension of the emergency room and enables ambulances to stay connected to the hospitals while transporting doctors virtually to the ambulance as well. That said, there are not many ambulances like this one in the world, and it is an expensive, complex project to implement in the rest of the countries.

On-demand healthcare solutions

Because society has become more mobile over the past ten years, more companies are keeping up with cutting-edge technology, particularly in the healthcare sector. Professionals are keener to work for several medical facilities at once rather than being tied to one employer.

For instance, on-demand healthcare apps allow doctors to offer patients so-called "on-demand" medical care, but only if a patient's needs align with their training, experience, and availability. As a result, doctors are better able to adapt their medical services to the changing demands of their patients.

On-demand healthcare encompasses a range of services, including scheduling follow-up meetings through mobile apps and conducting remote video consultations. Regardless of the use, the goal of on-demand is to enable patients to access these services quickly and conveniently, anytime and wherever they choose, using sophisticated mobile devices.

Patients now more than ever expect quick and easy access to healthcare. Consumer satisfaction with healthcare services depends on how quickly appointments are scheduled and how long the waits are. The determining factors are the convenience of appointment times and the convenience of the location or channel.

At the same time, as people have become more open to having smart technology play a more relevant part in their medical care, demand for digital health services has expanded quickly in recent years. Patients are likely to engage a virtual health coach or an intelligent clinician to assess health problems and discover the best treatment.

Consumers expect to utilize digital technologies to control when, where, and how they receive care services, affected by external experiences. By using digital technology in this way, healthcare will continue relying on them to enhance clinical expertise, free up clinician time, and personalize services, providing patients have more control.

Like most sectors, the pandemic pressured the healthcare industry to adopt new technology far more quickly than anticipated. On-demand healthcare has become essential during the pandemic. Many patients who are susceptible to or have had coronavirus symptoms have been forced to self-isolate at home due to lockdown limitations. They had little choice except to schedule remote sessions with their GPs via healthcare apps.

Patients can access therapy using on-demand mobile healthcare solutions without needing to schedule a doctor's appointment or go home. Patients can plan a call instead for a time and location that work with their hectic schedule. The patient experience has notably improved as a result. It's a lifesaver for people who would otherwise be required to spend hours each day traveling to appointments since they don't live close to the nearest practice or hospital.

Also, the service's quick, real-time nature makes it perfect for pressing matters. The ability to immediately schedule an appointment with a doctor on-demand allows patients suffering from acute symptoms, such as a severe headache or high temperature, to get help and guidance right away.

The days of providing healthcare only in person will soon be a thing of the past due to the rising popularity of on-demand consumer services and linked digital gadgets. Healthcare on demand is a reality. And it will be essential to provide the quick and simplified experience that patients of today want.

Telemedicine and virtual visits

The rise of virtual medical appointments is one of the most remarkable advances in healthcare. In comparison to an in-person hospital visit, it enables scheduling appointments with specialists at a time and location that is most convenient for you. After the COVID-19 pandemic breakout in 2020, telemedicine has grown even more popular, and most patients know how to use it.

In some hospitals across the United States, this digital transformation plan is already in practice, allowing patients to schedule virtual visits. They can send comprehensive prescriptions, set appointments, video chat with doctors on their phones or PCs, etc.

The increased use of telemedicine was driven by necessity but has created an enormous opportunity to reimagine virtual and hybrid healthcare models to enhance patient outcomes and access to care.

Not simply private practices or medical clinics are switching to telemedicine. Technology is being developed by hospitals to deliver cutting-edge care on a wide scale. The possibilities for offering cutting-edge healthcare remotely seem limitless, ranging from Bluetooth stethoscopes to remote cancer treatment via tele-oncology, among many others (Figure 10.2).

Figure 10.2 Telemedicine.

Transparent telehealth service delivery requires extensive change management. Healthcare firms may find this challenging since their fundamental business procedures have not changed in the last 70 years.

Patients prefer to be within easy reach of their treatment center. That is their most crucial selection criterion. Apps for telemedicine largely satisfy this need. It is not surprising that more patients and doctors are choosing mobile or online consultations over in-person visits, according to numerous research and polls.

Patients can now obtain immediate, on-demand care services without wasting time or resources due to virtual care and online patient visits. The standard of care shouldn't vary from regular in-person doctor visits. Video conferencing, live chat, cloud-based databases, and other communication methods help to assure this.

Manufacturers must pick the appropriate technology and provide adequate features for telemedicine apps to be readily and conveniently accessible. The processes that cannot be carried out digitally in the same way as they are in practice must be translated into digital form appropriately, or they may need to be revised. It's a common misconception that physical processes can be mapped perfectly. That is regrettably not the case. Instead, creating human-centered procedures for digital applications calls for specialized knowledge.

The cost of healthcare services has significantly decreased due to services for remote analysis and patient monitoring in real time. The expense of the apps is justified because of the favorable effects this cost reduction has had on patients, healthcare facilities, and health insurance. As more healthcare facilities join the telemedicine bandwagon and provide patients with medical apps, the overall cost per user for telemedicine apps will decrease.

In addition, telemedicine apps boost physician productivity while lowering the need for in-person visits, which boosts healthcare institutions' income. The importance of physical proximity is waning as well. That enables doctors to draw in and treat patients from a wider geographic area.

Apps for telemedicine ensure quicker and easier access to various professionals. Patients in rural and isolated areas with a shortage of specialists will find this very beneficial. However, appointments with the closest specialists are frequently not accessible for months, even in places with stronger structural foundations. The limited supply of "specialists" can now be used by society more effectively efficiently due to telemedicine: Doctors can refer patients to other specialists through apps.

Additionally, they can aid in locating and scheduling appointments with qualified professionals. Treatment by a specialist farther away is achievable with fewer issues due to the ability to span greater distances. That improves the workload distribution among specialists tremendously, raising the standard of care for all patients.

Chronic disease patients typically have to monitor their health data, goals, and treatments: Such apps can help patients in this particular category.

Patients' health improves as a result, and health insurance expenses go down. Thus, involving patients in their care and treatment leads to higher-quality and more affordable care.

Medical technology is rapidly improving due to the digital transition, which also creates new opportunities. New digital ideas must not only be clever but also human-centered if they are to provide meaningful value to patients and doctors. We have more than 20 years of experience guiding businesses toward new services and goods. We have been creating medical technology solutions as digital experience designers for a while now, ensuring improved customer satisfaction and product success via human-centered design.

We anticipate a sharp growth in the number of telemedicine solutions in the upcoming years. In the upcoming years, telemedicine will become the new standard for patient care and treatment. Considering that telemedicine is a legitimate medical practice. Instead, it flawlessly exemplifies the digital revolution of healthcare that the modern healthcare business hopes to achieve in the upcoming years.

Patient portals

A fantastic trend in the evolution of healthcare solutions is the development of particular healthcare platforms where patients can examine their medical records, check their prescriptions, make an appointment with the experts, and consult with their doctors or request more information from them. Also, they can obtain the laboratory findings and give the medical professionals their health information.

And they are just a handful of the alternatives people have access to on healthcare platforms. In addition, this system lessens the workload for the medical staff by enabling quicker and more access to EHRs.

A website, which is online connected to the EHR, is referred to as a patient portal. That simplifies access to all patient data. In other words, a healthcare portal gives you access to numerous data kinds, like lab results, family histories, summaries, or immunization history. You ought to get all of this information from a perfect portal.

Functions, such as online appointment scheduling, secure bill payment, and direct secure communications, are offered by online patient portal software. Many of them, meanwhile, lack such detailed information. The information accessible through these portals is based on the standards of the particular healthcare family and the portal vendor.

Nowadays, there are more and more people using medical portal websites. Patients can access their health information at the hospital, and many portals allow for safe texting.

The patient portals can be accessible on mobile devices and tablets in addition to PCs. Healthcare professionals use mobile-friendly patient portals to communicate with their patients. As they employ these tools to improve the portal by raising patient involvement, the web portal development business plays a significant part in assisting medical professionals in providing care outside their clinics.

Patients become more aware of their health when being able to access their medical records. Therefore, they are better equipped to communicate with their healthcare professionals and comprehend their conditions.

Patient interaction tools increase loyalty. Many elements, such as encrypted texting, help form a close relationship between patients and their doctors. This familiarity with their needs leads patients to return to the same expert again. Having medical information close keeps doctors mentally prepared and allows them to act quickly if necessary, saving patients from chronic illnesses.

Patients can use the portal to finish tasks that ordinarily involve one or more phone calls. The ability for patients to seek appointments, referrals, and prescriptions straight from the portal improves clinical staff's productivity and allows for patient assistance in situations where you require urgent care and questions.

The creation of the patient portal makes it possible to electronically complete the registration forms before check-in, maintaining the efficiency of the office. This entails that the front office employees can concentrate on the patient and provide the information they require, saving them the trouble of walking and ensuring they are comfortable before seeing the doctor. Ultimately, it results in cost savings to run the business and provide better care.

The patient–physician interaction is now more intimate than ever because of patient portals that give patients access to their providers round-the-clock to evaluate patient health information, ask and receive answers to queries, and read notes.

The patient portal helps with chores like prescription refills and referral requests, making patients more relevant. Clinical outcomes increase when patients adhere to doctors' orders.

With the help of personnel, actions that patients would typically perform manually can now be done electronically due to the patient portal. By giving patients online access, notifying them of the need for refills and responding to inquiries about referrals, staff members can avoid appointments.

Patient portals are currently present in most medical institutions, so the next step is to expand these systems to create facilities that can accommodate more patients. It has historically been impossible for many patients to generate sign-up screens due to the clumsy and inconsistent user experience. However, the initial exhilaration of patients driven to use the self-service portal when the counter UI fails is followed by frustration and resignation.

Patients who have had disappointing experiences using the patient portal may find it challenging to use it in the future. Because of this, UX optimization ought to be one of the first upgrades to be considered for any current patient self-care system. As more people use mobile devices to manage their health, the broader shift toward mobile devices impacts the healthcare industry's digital transformation.

A mobile version of customized patient portals enables users to navigate their health journeys. Also, smaller healthcare organizations are utilizing pre-built apps and extending them to offer patient self-care features on mobile devices. Before their first consultation, patients can register more quickly and easily by filling out digital patient registration forms with their information and consent.

The digital form is a simple way to promote patient communication by lowering patient wait times in medical facilities and allowing secure end-to-end data control. Speeding patient flow, lowering the risk of clinical errors, minimizing front-desk workload, and providing a comprehensive perspective of the patient medical history are the key factors of the healthcare transformation.

Online registration is already supported by a large number of patient portals in use. More importantly, you can add more functionality to them or connect off-the-shelf solutions with others. Because many patients complain that these systems are too complicated and how difficult it is to understand and use their interface features, this digital revolution in healthcare is still in its early phase.

Health wearables

People today are more concerned about their health than ever before; rather than going to the doctor when they are ill, they constantly search for efficient yet practical solutions to check their health indicators.

In essence, this was the main factor for the sharp rise in wearable medical device sales. The digitalization of healthcare makes it possible to track several health variables and deliver precise health data in real time.

Heart rate monitors, exercise and fit (e.g., duration, type of activity, distance, calories burned, etc.), sweat meters (e.g., for measuring blood sugar – a crucial routine of diabetics), and oximeters are just some of the popular types of health equipment.

Apple Watch is one of the most well-known wearables to hit the market. This smartwatch can record body temperature, weight, and periods, and monitor heart rate and exercise. Apple Watch can even send you reminders to drink water or wash your hands. Doctors can then utilize this information to analyze health parameters, make diagnosis, etc.

The epidemic made smartwatches more useful for keeping track of health. As COVID-19 spread, blood oxygen saturation (SpO2) smartwatches became widely available, alerting users to low SpO2, a life-threatening condition – challenging for people to recognize on their own. More than 10% of US consumers now use their smartwatches to look for COVID-19 symptoms. Given that 15% of US consumers owning smartwatches bought them after

Figure 10.3 Smart device

the start of COVID-19, the pandemic may have even boosted sales of the devices, according to a research by Deloitte (Figure 10.3).

The development of smartwatches is advancing quickly due to developments in sensors, transistors, and AI. As an illustration, several smartwatches now come equipped with optical sensors that continuously track changes in blood volume and composition using a method known as photoplethysmography (PPG). This data is being used by machine learning to create and continuously improve algorithms that offer insights into users' activity levels, stress levels, abnormal cardiac patterns, etc.

Another illustration is the development of blood pressure monitoring capabilities for smartwatches employing PPG and other technologies like Raman spectroscopy and infrared spectrophotometers. Cuff blood pressure monitoring is cumbersome and painful. Furthermore, routine blood pressure checks may not detect chronic hypertension, a risk factor for heart disease, heart attacks, and strokes. The smartwatch industry could grow if blood pressure could be measured accurately, continuously, and unobtrusively: 1.3 billion adults have hypertension worldwide.

The capabilities of the existing smartwatch sensor technologies are, of course, constrained by their inability to adhere to or penetrate human skin. Smart patches can help in this situation.

Smart patches are often small and undetectable, adhering directly to the skin, and created by MedTech businesses. Some "minimally invasive" smart patches function as biosensors and occasionally dispense drugs using minuscule needles that gently enter the skin.

Primarily, smart patches are made for a specific indication, such as the management of diabetes, patient monitoring, and medicine delivery, in contrast to smartwatches which offer a wide range of health data and insights. Additionally, a wider range of technologies is used by smart patches.

For instance, ECG technology, which records the heart's electrical activity directly and more precisely than smartwatches, is frequently used in smart patches that detect heart rate variability.

Smartphones and smartwatches continue to be quite significant. Smartwatch and smartphone apps integrate data from smart patches to send data to these devices for display and analysis. Doctors might view wearable health data on a patient's health record with the correct technology, including interoperability capabilities and gain access to more comprehensive information to help with diagnosis and treatment.

Disease history analysis

More tools are available nowadays for examining a patient's disease history and providing recommendations to clinicians regarding treatment. The business BostonGene is an example of such a solution. The BostonGene algorithm thoroughly studies all patients' health issues and provides a customized treatment plan that may result in more valuable outcomes.

Drug research for rare diseases is expanding, although mostly there are no approved treatments for most of these illnesses, which are frequently poorly understood by the scientific community.

Measuring the right objectives and successful recruitment through choosing clinical locations and patient identification, and optimizing rare disease interventional trials early in clinical development contribute to avoiding trial errors. Studies on the natural history of diseases can serve as the basis for optimal interventional trials, particularly in the case of rare diseases.

Several factors of conducting an early natural history of disease study are incredibly relevant:

- Characterizing a rare disease or rare subtypes,
- Planning the interventional trial and establishing trial endpoints,
- Building relationships with sites and stakeholders,
- Engaging patients before the interventional trial recruitment,
- Leveraging operational efficiencies.

Although value can be achieved throughout the clinical program, the following advantages are enhanced by conducting natural history research as early as possible in developing rare illness drugs. Using end-to-end solutions and scientific and methodological knowledge help optimize this value.

First, with future interventional trials in mind, the medical team contributes to the protocol and study design. Working with patients or a group that advocates for them is highly advised during the study planning.

Next, a variety of data-collecting techniques must be taken into account. In rare diseases, it's crucial to employ enriched or secondary data to increase the number of eligible patients and lighten the load on treatment facilities and patients. Then, to plan interventional trials and have regulatory talks, it is necessary to ensure prompt reporting of important study data.

To organize the engagements and seek alignment early in clinical development, the medical experience should be leveraged to involve regulators in the natural history of illness investigations. This is especially pertinent if a single-armed interventional trial is being compared to the longitudinal dataset from the natural history study as an external or historical comparator.

The results of the natural history study's operations and data analysis can be used to improve interventional trials. Sponsors can avoid interventional trial mistakes, such as choosing the incorrect endpoints, down the road, by asking the proper questions early in clinical research and engaging with teams who comprehend the specific demands inside the rare disease sector.

IMPROVING PATIENT EXPERIENCE

Digital innovation is the most crucial and beneficial strategy to address the need for expansion in health systems. But the goal goes much beyond merely updating the infrastructure. From a consumer standpoint, digital innovation must aim to transform the healthcare business model.

Nearly all growth pathways revolve around digital. It is crucial for demand generation, aggregation, and capture but is also helpful for increasing lifetime or long-term customer value capture, which changes the customer acquisition cost (CAC) vs. lifetime (LTV) equation.

Other industries are already witnessing the results of a digitally driven paradigm. The way we look for and book travel has changed because of companies like Expedia and Travelocity, which also shed a light on untapped industry insights. Banks were able to increase the lifetime value by learning more about their customers and using that knowledge to create customized offers.

These same concepts must begin to guide and be adopted by health systems. Customers now have mission-critical needs that must be met if they are to be able to easily access, explore, and book services and care alternatives across all channels. Consumer expectations have altered across almost all other industries, including financial services, dining, travel, and retail. No longer can healthcare remain the exception.

A deeper awareness of consumers in the market across different lines of care is delivered through insights from customer data platforms and identity-driven engagement, enabling marketing to be much more efficient and tailored.

Health systems must improve their ability to measure effectiveness to implement this digitally driven strategy. The number of patient or customer accounts, also referred to as Digitally Registered Users (DRU), and metrics like Monthly Active Users (MAU) and DRU give insight into how the consumer funnel actually functions. Additionally, the creation of Return on

Marketing Investment (ROMI) models aids in transforming marketing from an expense to an investment that supports expansion and may be linked to particular campaigns.

In recent years, funding for health technology businesses has been aggressive and record-breaking. Health systems have a huge chance to take advantage of the innovation market and funding by collaborating in ways that enable the development of new products and services, which will open up completely new revenue streams.

The healthcare industry in the United States is highly specialized and complex. Products and solutions that have the flexibility to support the ecosystem have an edge. Imagine a streamlined and uniform platform that provides a comprehensive perspective of a patient and streamlines interactions with patients in terms of transactions, communications, and engagement, while also making the best use of limited resources and preventing caregiver fatigue.

These platforms give health systems the ability to fully capitalize on market investment and speed to drive transformation. Businesses that focus on being valuable platforms, taking into account the diverse spectrum of the operational and business logic of various constituents without pressuring organizations into a particular operating model, will succeed in this market.

This subtle method of producing commercial goods has the potential to revolutionize the game. It's also why digital advances created within a health system may end up being the most beneficial overall.

Beyond encouraging client acquisition and retention, digital transformation has the potential to alter the health system's economics to support brand-new business models that foster expansion. Digital not only makes it possible for solutions to be scaled up, but it also makes it possible for more meaningful and direct consumer engagements that can have a bigger influence on businesses.

Think about how health institutions can divide their clientele based on patient groups, diseases, and needs while still providing upscale services that could assist patients between treatment periods. There are numerous opportunities, including brand-new business models, direct-to-consumer or direct-to-employer products, and even new varieties of insurance firms. All of these digitally centered models have the potential to significantly alter the way healthcare systems operate.

Growth strategy and digital strategy cannot be discussed independently. Compared to "average" hospitals, hospitals that provide "excellent" customer service to their patients experience much higher net margins. Digital will continue to be the primary force behind and provider of those experiences.

Digital cannot be seen as an addition to current healthcare systems like a veneer. To generate self-disruption of our whole business model and promote sustainable long-term growth and recovery for our organizations, it is essential to recognize digital as an organizational value driver.

BENEFITS OF DIGITAL TRANSFORMATION IN HEALTHCARE

In truth, there are several advantages of medical digital transformation for both patients and the healthcare organizations who use them. By utilizing these advancements, medical professionals and hospitals can streamline their operations, obtain more precise patient data and health indicators, and develop a more effective treatment plan faster. Of course, all of these elements positively impact the outcome.

Many firms and sectors underwent a digital transition over the past few years. Whether they want it or not, this trend is revolutionizing various industries and changing how business is conducted. Like all other data-driven businesses, the health sector is rapidly growing and adapting to change.

But this procedure introduces several new difficulties and issues that this sector has never had to deal with. Making an efficient data pipeline that will enable healthcare institutions to share and manage data inside without harming the expansion and efficiency of this sector is a major problem. Digital technologies are influencing change and transforming the healthcare industry, enabling firms to provide individualized, excellent patient care and sustainable growth.

Healthcare organizations can benefit from digital transformation by developing integrated systems and procedures that benefit patients and medical personnel and allow them to deliver the required care and services with more effectiveness and precision. Let's examine how the healthcare sector, patients, personnel, and services can profit from the most cutting-edge sophisticated digital solutions in light of all this.

Consistent hospital and patient data management

Every day, hospitals must work with enormous amounts of data. Every day, they must be able to provide, compile, and maintain medical reports, bills, and other patient data. It would take a lot of time and effort to do this accurately by hand.

However, with digitalization, all these hospitals and medical procedures can become more effective, seamless, and quick. By going digital, the entire workflow can improve. Effective and reliable data management is the main obstacle to digitization, and modern hospitals and healthcare organizations that are open to it must find a solution.

These organizations must expedite the procedures to provide better patient care. The ideal options for medical staff and doctors to obtain, exchange, manage, and save data include centralizing data storage, workflows, and intranets.

Connection across social media

One of the best things about digital transformation is that it enables businesses to communicate digitally with people via the numerous social media platforms that are currently accessible. Through this connection, medical organizations respond quickly to patients' questions about various illnesses and problems.

Through internet-enabled mobile devices, patients can post their questions at any time, anywhere, and get qualified solutions in real time. Patients can access all the information they require via mobile devices instead of physically visiting the hospital. Patients will receive the prompt treatment they need due to the rapid service.

Aside from that, it also increases confidence in the patients and fosters trust. Hopefully, people will be more willing to seek assistance when they need it rather than delaying requests until their situation worsens. Additionally, it can aid in educating patients on numerous topics about their health, well-being, and medical care.

Improved internal communication

Communication is essential for the healthcare sector as a whole to run well and give patients the treatment they need. Healthcare organizations rely on contemporary technology much like the rest of the world does. These organizations need digital solutions to enable more effective and seamless communication across institutions.

Examples of digital solutions are websites, message boards, VR equipment, films, etc. These tools enable healthcare providers to communicate with one another and their patients and improve the quality of their treatments, patient care, medical advice, and everything in between.

Electronic health records

Healthcare companies should get rid of paper records and organize their daily processes and procedures with digital health records. The first and most obvious advantage of adopting digital is that human error is virtually eliminated.

EHR is a great way to increase productivity and is also very beneficial for both practitioners and patients. These records assist medical professionals in accurately diagnosing patients' problems and delivering timely care.

This might, in some circumstances, be the difference between saving someone's life. Emergency care providers particularly benefit from EHRs. Every second matters when treating critically ill patients, and EHR enables them to get patient data much more quickly than before and deliver essential care and appropriate and prompt treatment.

Data analysis

Healthcare companies are gathering massive volumes of data. To maximize the value of this data, they are utilizing big data, AI, and Internet of Things

(IoT) technology. Increased data gathering and analysis bring innumerable advantages, such as better disease detection and treatment, disease prevention, and patient-specific personalization of healthcare services.

These are the three ways healthcare organizations are leveraging technology to enhance data analysis:

1. **Cloud:** Data exchange, more convenient search and retrieval, and interoperability between medical systems, IoT devices, and applications are all made possible by cloud systems.
2. **Artificial Intelligence (AI):** In addition to processing massive datasets to identify trends and forecasts, it may be used to evaluate medical imaging and EHR data for diagnosis, taking the place of human experts in many fields.
3. **Big Data Analytics:** It is used to track information important to healthcare organizations, such as reports of disease outbreaks, complaints, and reviews of healthcare services, in public health records, social media, and other data sources.

APIs and interoperability

To digitize medical treatment, data must be exchanged securely and efficiently. The proper data interchange between EHR systems, medical equipment, and other integrated services must be ensured using application programming interfaces (APIs). In the end, this may result in a diagnosis and treatment that is more precise and timely.

Organizations must also think about security when developing APIs. They can prepare better to offer cutting-edge digital services if they invest more money in safe, legal system connections.

Healthcare mobile apps

The burdensome task of helping patients locate nearby doctors and make travel arrangements to hospitals is lessened by the appearance of healthcare mobile applications. Through healthcare applications, practitioners and doctors may engage with people and provide helpful consultations, counsel, and care.

It's crucial for people throughout the world, but especially in remote areas where access to medical facilities is constrained. On the other hand, patients can utilize healthcare apps to quickly and easily look up medicine information, research conditions, ask questions about the topics that interest them, receive helpful advice, and much more.

Whatever perspective you pick, it is clear that digital transformation is advantageous for the healthcare sector as a whole. Modern digital healthcare platforms have been developed as a result, which has increased operational

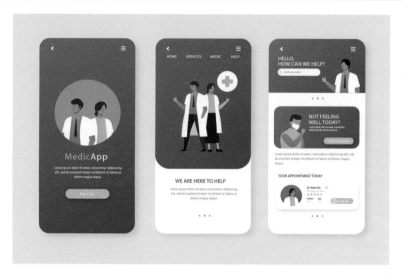

Figure 10.4 Example of the Health application.

effectiveness and enabled healthcare institutions to provide patients with an integrated care plan (Figure 10.4).

The world's industries, including the healthcare sector, are significantly impacted by modern technology. Due to this beneficial effect, contemporary healthcare technology-driven solutions, such as blockchain EHRs, AI-enabled medical devices, and telemedicine, have been developed.

LIMITATIONS OF DIGITAL TRANSFORMATION

Although there is much potential for digital change in healthcare, there are also many difficulties. The healthcare sector is facing several significant obstacles that turn service digitization into a challenge.

First, it's crucial to realize that cybersecurity concerns link to technology usage. Cybersecurity experts are essential to ensuring the safety of patient data and preventing any data loss. There is a severe lack of individuals with comparable credentials in the healthcare industry today. The lack of skilled workers may increase risks and expose system weaknesses.

You should be aware that most healthcare providers still rely on their legacy systems, which link to a wide range of security risks, in addition to the lack of personnel who will integrate digital innovation on a quality level. Even if those systems carry out all of the essential activities, many institutions are unwilling to update them, and maintaining them is difficult. As a result, they delay technological change and impede the adoption of new software.

Data management

Healthcare faces significant challenges with data processing and analysis. The vast amount of data that hospitals, clinics, and healthcare professionals gather contributes to the issue. Better individualized treatment is impossible for companies to give without an AI system that can analyze this data.

A further significant difficulty is in gathering and synchronizing data. It is challenging for medical practitioners using telemedicine to update patient health records while operating from various locations and platforms. Health records must be able to be recorded and updated from different devices during in-person and online visits.

Cybersecurity

For healthcare institutions, one of the most crucial jobs is to address the risk of cyberattacks. Healthcare is one of the top sectors that cybercriminals target, according to numerous security research reports. The volume and accessibility of patient and medical data, the high number of IoMT-connected devices, and the complexity of patient care delivery models all contribute to rising risks.

Healthcare businesses are increasingly storing data on the cloud to foster collaboration and data sharing, which presents a problem for cloud security.

The attack surface of the healthcare security perimeter, which includes many organizations, IoMT devices, cloud systems, and conventional medical equipment, is substantially broader. Additionally, many firms lack security protections despite increased security supervision and investments.

Most hospitals keep tens of thousands of medical devices without the necessary security precautions. Despite firms' dedication to digital change, risk management procedures may lag.

Data privacy

Patient data exposure is becoming an issue as the healthcare sector transitions to collaborative healthcare. A collaborative medicine approach makes patient data accessible across intricate medical settings, allowing many users to access it from many devices and locations.

Medical personnel and biological researchers also must have access to data to improve medical judgment and patient care. That includes using AI for health monitoring and diagnosis. Without appropriate data privacy safeguards, there is a high risk of losing personal information and having medical treatment compromised.

DIGITAL HEALTH: THE FIRST INNOVATION IN HEALTHCARE THAT CAN REDUCE COSTS IN THE LONG TERM

The smartphone counts steps, the Fitbit fitness band records calories burned, heart rate and workout progress, the app mySugr uses exercise data and food diaries to predict blood glucose curves for diabetics, and the app Tinnitracks plays music for tinnitus sufferers that filters out certain frequencies to neutralize the hearing disorder.

Digital solutions for sick and healthy people have been around for a long time, and they are being eagerly downloaded. According to a leading e-health publisher, more than 50 percent of US cell phone users have downloaded a health-related mobile app.

Healthcare institutions have been slower to embrace the possibilities of digitization than patients themselves. Yet they could benefit significantly. In Sweden, for example, dedicated digitization of the value chain has led to significant savings in the healthcare system. In 10 years, gross savings of up to 25 percent are possible. While more is needed to reduce spending, it would significantly mitigate the expected cost increases.

Digitization of the Swedish healthcare system using state-of-the-art e-health technologies could bring gross savings of 25 percent by 2025

Digitization offers an exceptionally high potential for healthcare. Ubiquitous sensors combined with apps and systems that rely on human-centered design and intelligent algorithms (advanced analytics) to analyze the volumes of data generated could revolutionize medicine and wellness as a whole if the system was not afflicted with its inherent braking mechanisms.

Take, for example, MyTherapy, an app that helps patients take their prescription medications – which drugs, how much, when, and how often. In particular, older patients with multiple illnesses must take many medicines simultaneously. Numerous studies have shown that very few patients can do this. According to an estimate by the World Health Organization, medication compliance in industrialized countries is only 50 percent for long-term treatment. In the United States, medication non-adherence causes nearly $300 billion in additional healthcare expenditures annually.

In the digital world, mobile solutions address this very problem. With the MyTherapy app, patients can scan the barcodes on the packaging with their smartphone and receive a notification at the appropriate times after entering the prescribed intake intervals. Game elements in the app help the user to keep track. In this way, not only the patient but also the healthcare system benefits from better compliance, as secondary problems caused by non-compliance do not have to be treated.

Lack of a suitable business model

Although such an app is an excellent idea, a suitable business model is still not available. In many countries, citizens receive free access to primary healthcare and are willing to pay 99 cents or less to download the app. Health insurers require clinical studies on efficacy before considering covering the costs. The pharmaceutical industry invests millions in such studies, but this is impossible for a start-up. Doctors have little motivation to recommend such apps to their patients. This may be related to their unfamiliarity with the various apps or needing more confidence in the app's clinical relevance or quality. Questions of reliability or data protection almost always play a subordinate role. In addition, a physician who recommends such an app does not immediately benefit financially and may have to spend a lot of time explaining it. This leads to a dilemma: Although there is proven added value for digital health at the system level, no one wants to pay for it. The patient has already paid for healthcare with their premiums, the health insurer is only willing to pay for proven individual added value, and the physician wants to be the one to invest time only if they are compensated for it.

Remuneration is based on fee-for-service models, where individual services are billed. The quality of the treatment and the outcome's success is not included in the remuneration. Thus, the basic mechanisms of the healthcare market in many countries stand in the way of the widespread adoption of digitization. The United States, which is not exactly a role model for healthcare costs, is now taking a different approach: the state health insurers Medicaid and Medicare are leading the way by including a quality component for hospitals in their reimbursement models.

Apple is taking a twin-track approach with its ResearchKit and CareKit platforms. ResearchKit is an open-source software platform that invites physicians and researchers to build apps. Volunteers will download these research apps – which, if permitted, can also access other health apps on the smartphone – to their iPhone, allowing data to be collected for medical research. Apple hopes that, by doing this, its iPhone will become a tool for medical research. ResearchKit enables medical researchers to investigate various conditions like asthma, breast cancer, and Parkinson's disease. Top US universities and research institutes are on board with this project.

CareKit is aimed at the patients themselves. The platform is designed to help patients manage their conditions better while also allowing their vital signs to be shared with their treating physician. In early 2016, Apple launched four self-developed modules: Care Card, which reminds users when it's time to take their medicine or exercise; Insight Dashboard, which records symptoms and relates them to the measures from the Care Card module; an app that keeps track of mental health; and Connect, which sends the data to the physician or a family member. Apple cannot access individual data and has committed to respecting and protecting privacy.

CareKit is also an open-source platform – partners are invited to develop apps on this platform to extend the range of services. With the acquisition of Gliimpse, a personalized health data collection/dissemination platform in 2016, Apple is well positioned to become a key enabler for patients to more directly influence and redirect health decisions.

Naturally, the established players in healthcare technology are also attempting to develop a platform and secure a central position in a new medical ecosystem.

Philips is bundling its business as Philips Healthcare with a product range from toothbrushes to magnetic resonance imaging (MRI) scanners. In addition to the hardware, it has also developed software for managing entire hospitals. General Electric operates its Health Cloud, which sells server capacity and rents software for the healthcare system (software as a service).

IBM acquired Truven, Explorys, and Phytel, companies that have collected vast amounts of health data over the years, and uses this wealth of data as the basis for solutions in population health management.

Microsoft has launched its health cloud boosted by Cortana, its proprietary AI capability, to become the preferred platform for developing digital health solutions. Siemens Healthineers has a clear digitization strategy with Teamplay, a cloud-based digital solutions platform used in many hospitals.

PREDICTIVE MAINTENANCE FOR PEOPLE

Although the players in the digital healthcare industry may still be founding their place, there is no lack of vision. In the United States, pioneers are working on population health management. The idea mirrors what machinery manufacturers have already achieved with digitization: predictive maintenance, where a part is repaired in time before it fails. To transfer this concept to people, volunteers send the health data collected via their health and fitness apps to a central body, which then evaluates it.

If worrying deviations from the norm are detected – for example, if the user is increasingly overweight with rising blood pressure – the program responds and recommends targeted exercise regimes and a nutrition plan – all via an app that also records the impact and reminds the user if a unit has been skipped.

The tailored program even includes a voice assistant with AI that acts as a personal health coach and can answer questions. This predictive maintenance improves the user's health before they become ill.

The question isn't whether we will use digital health services in the future but when and who will provide and regulate these services. Existing healthcare organizations like the large health insurance companies in the United States and the British NHS still have an invaluable lead over Google and Apple with their access to highly standardized and granular patient data. This lead can be defended only by healthcare systems open to innovation and willing to shape it. The window where this must be done is already available, and waiting is not an option. Otherwise, there is a risk that, just as in other industries – the digital champions won't be the incumbents.

CONCLUSION

Every industry is impacted by digital transformation, and healthcare will surely see more of it in the future. However, the industry may be holding back from fully embracing digital transformation due to high risks and fear of modernization.

Cybersecurity issues are relevant to every new internet-connected technology. To maintain the protection of patient data, an organization's cybersecurity personnel must be accessible to help with adoption and implementation.

According to a report by ISC, even though the worldwide workforce must increase by 65 percent to adequately defend critical assets, the cybersecurity workforce gap has closed for the second year in a row. Because of the labor shortage, the remaining personnel can be overworked, resulting in serious security flaws and shady network behavior being unnoticed.

Healthcare relies largely on outdated systems to carry out crucial activities in addition to having a shortage of individuals willing to dedicate themselves to digital transformation and implementation. Because of their portability and difficulty in patching, legacy medical devices raise security issues. Additionally, companies might not want to take the chance of updating these devices while they are still providing essential services. Healthcare businesses must address a wide range of clinical requirements and applications.

Organizations must prioritize the privacy and security of PHI, whether using EHRs or AI-powered analytics solutions (PHI). These new technologies might also call for greater employee training and security precautions, adding to the workload.

When it comes to digital transformation, healthcare faces particular difficulties. However, that doesn't mean that the sector is sailing unknown waters. It can strengthen its digital transformation efforts by studying the achievements and pitfalls of other industries.

Unfortunately, there is no quick fix for the industry's pervasive reliance on antiquated technologies and the shortage of cybersecurity workers. These obstacles are here to stay, but secure digital transformation in healthcare is not impossible.

Reducing some of the workloads on the personnel and relying on technology to fill the manpower gap can help reduce workloads and enable safe and quick technological adoption. Industry clouds, for instance, might encourage digital transformation. Industry clouds provide sector-specific cloud solutions that support enterprises' efforts to streamline operations and eliminate waste.

Such solutions are becoming more prevalent across all sectors and are constantly changing to take into account brand-new industrial difficulties and combine cutting-edge digital capabilities. Investing in integrated solutions that enable automation can also help allay security and privacy worries while freeing up staff to focus on additional improvements to the security architecture of their company.

Finally, businesses need to recognize when to work with subject-matter specialists to lessen resistance to modernization. Organizations are unable to transform their own. Because mistakes can be expensive, they need to partner successfully with individuals who can offer that guidance and create a roadmap for digital transformation.

Chapter 11

Digital transformation in the legal industry

Even the most ingenious inventions in the legal industry are useless if lawyers lack the skills to use them. Lawyers must first consider investing in talent and people who can make that technology effective if they want to fully realize the potential of digital transformation in the legal industry.

Diverse businesses and sectors are experiencing the emergence of new business models that are changing old value chains and dissolving sectoral silos. Additionally, we have seen how unanticipated rivals – startups and even businesses from unrelated industries – have been disrupted by fundamentally altering how they were doing things (Figure 11.1).

The law industry has experienced the same thing. Legal professionals can now lead the way in enabling more secure, effective, accessible, and client-centric legal services based on digital transformation solutions thanks to disruptive technology. Additionally, firms are reinventing themselves by turning data into valuable assets.

Lawyers must understand how technology affects various industries and consider how law, business, and technology are intertwined to compete in this new environment. Lawyers can devote more time and effort to client service and higher-value work thanks to the effectiveness and support of new technology and operating systems.

However, individuals with a history of success following "what has always been done", find it particularly challenging to understand the concept of actual disruption. This is true for most large legal firms and organizations, where there is a conflict between lawyers who want to protect the "profession" and consumers' requirements and an industry that wants to provide those needs.

However, the epidemic has brought those resistant to change to a turning point where adopting technology is no longer a choice but a requirement. Businesses and law firms that extensively rely on technological inputs and facilitators have realized the value of disruption.

Smart businesses have been refining internal procedures, rewarding efficiency, and exploring new business models that maximize customer value.

Will legal firms and internal departments need to reevaluate and alter their long-term business strategies in this new post-Covid environment?

DOI: 10.1201/9781003305163-11

Figure 11.1 Digital transformation in legal industry.

Will business people and leaders embrace technology, recognize, and develop new market opportunities, and still act quickly and aggressively to address upcoming challenges?

Diversity, teamwork, and the organization's capacity for cultural adaptation are all impacted by digital change. The legal sector must do a better job of including all groups, including lawyers, engineers, data analysts, and persons from different racial, ethnic, and gender backgrounds.

In terms of talent, the sector needs to be open to new paradigms; additionally, for law firms and businesses to succeed in the age of digital transformation, up-skilling and re-skilling lawyers in new technologies is crucial.

Lawyers must establish a reputation for being change enablers in the increasingly complicated and dynamic digital economy. To achieve the results that clients desire, lawyers must reengineer themselves. A new market for legal services may become available due to technology, substantially boosting service quality, speed, and accessibility while reducing costs and inefficiencies at the workplace.

So why do many legal firms still operate on antiquated technology and a sea of paper?

The basis of how legal procedures operate may hold the key to the solution. Protecting client confidentiality, exercising restraint, and relying on precedent are all requirements of the legal profession. The delayed adoption of new technologies by law firms, although doing so would undoubtedly benefit both staff and clients, may not come as a surprise after all.

People can now manage their legal company like a real business and use their time more effectively due to digital transformation. Law firms must develop a plan and begin putting it into action immediately if they don't want to fall behind in a few years.

The legal industry is embracing digital transformation due to the need to remain competitive. The increasing volume of work, clerical errors, lost personal time, and burnout has been common pain points in the legal sector. Digital transformation could be the solution all legal firms need to become more efficient and profitable.

Digital transformation looks at quality and client experience as central principles. Having a high-quality digital law firm doesn't require only having the right software. The primary force behind the digital strategy should be market research to identify the target audience.

Two key factors prospective clients look for are simplicity of use and superior security. The bottom line is security – the need for everything to be secure and encrypted. The businesses should adopt a "zero trust" security strategy, mandating two-factor authentication and authenticator applications for all interactions, even editing the website.

Presentation is also a crucial component of attracting and satisfying distant clients. The infrastructure that improved video meetings can be held with, such as superb microphones, decent lighting, and strong internet connections, should be invested in by law companies. Professional video and audio are crucial for virtual meetings with clients. Since the majority of firms are moving online, attorneys will need to be able to establish credibility and dazzle on camera.

The technology that enables the remote operations of a company also decreases its costs. Documents that would have previously been delivered back and forth by international couriers can now be signed digitally by lawyers. The additional time and expense required to pay someone to perform potentially automated duties, something law firms frequently undertake out of pure inertia, is another element to consider.

Client expectations, however, are what will eventually fuel digital transformation. Clients will eventually object when charged for writing a letter that could have been completed in two hours with automation and only required a fast inspection. You can capture so much more value by embracing digital transformation, and the clients benefit from it as well.

That said, digital transformation is a journey rather than a final destination. Clients' expectations are constantly going to change. The last thing you want when they work with you is for them to have a sense of time travel.

DIGITAL TRANSFORMATION IN LEGAL FIRMS

Law firms pride themselves on putting clients first. Most firms' employees, including the managing partners, client-facing attorneys, and numerous support workers, have functioned outside the technological realm. Except for email, customer databases, and online document signing, the legal profession has been sluggish in embracing digital transformation.

What does digital transformation look like in legal firms? The process of implementing technology across the entire organization to increase efficiency, decrease errors, expand capabilities, deepen insights, and strengthen customer relationships is known as digital transformation.

In furthering corporate digital transformation, which has as its end goal value creation and improved consumer experiences, legal has a vital but underappreciated role. Like technology, the legal function now influences every aspect of business and is no longer vertical.

Legal must operate as a proactive, data-driven, comprehensive risk mitigation, agile business unit instead of as an exclusive, self-contained department. This holds for both the outside resources and the corporate legal department. The legal department's services to the business and its clients must go far beyond providing legal counsel. It is necessary to reconsider its goal from both the corporate and customer viewpoints.

Scalability presents an opportunity for investors, business- and tech-savvy suppliers, and customers as the industry transforms from practice to product and labor intensity to tech efficiency. There is a distinct difference between digital updating, based on software investments, and digital transformation, which involves updating technology and internal working methods at all levels and customer interactions by the added value.

This implies the creation of fluid and collaborative structures where work is based on knowledge management (KM) through new management models. There, control must be transferred to new models of co-creative leadership that enhance the capacities of everyone in the office, avoiding underusing personal and material resources.

DIGITAL TRANSFORMATION TRENDS IN THE LEGAL INDUSTRY

Varied legal firms have different approaches to digital transformation. Each law firm's area of expertise and particular clientele determine the digital transformation type that is most important to them. Corporate law companies, for instance, will have different standards than family offices. The company must understand all its possibilities before concentrating on those that could positively impact its business.

Knowledge and information are crucial to the performance of legal services, and confidentiality and privacy have long been crucial aspects of the attorney–client relationship. As a result, confidential communications and information security are highly valued in the legal profession.

Globally, the adoption of technology has undergone considerable change as a result of the lockdown and remote-work trends of the previous two years. Having a clear digital strategy is essential for future success, but just 23% of law firms claim to have one in place, according to a PwC survey.

Clients want their service providers, including the legal industry, to stay up with trends that facilitate communication and outcomes as technology is rapidly adopted by consumers.

Automation

Technology is developing to remove these tedious chores from a lawyer's day, such as submitting paperwork, signing contracts, and checking documents for mistakes or discrepancies. Lawyers spend countless hours reading and examining documents and contracts to find inconsistencies or prepare an argument. Technology can support these procedures and enhance insights and accuracy while easing part of the workload for attorneys and subordinate staff.

To maximize their investment, today's clientele is technologically savvy and demands the same of those who deliver the services they hire. Lawyers are becoming more and more aware that the current paper-based system is outdated and is virtually likely not sustainable in the long run. Technology can change how attorneys practice law, how they provide legal services, and how people can access the legal system and judicial systems.

Legal professionals can shorten the time it takes to complete duties like form filling and document filing by using digital process automation solutions. They can also aid attorneys in keeping track of their cases' development and effectively managing their case files. This software family aids attorneys in the management of electronic evidence, such as client communications, exchanged documents, recordings, pictures, and videos. It can aid lawyers in swiftly and effectively reviewing, exchanging, debating, and analyzing this material.

Law companies are not an exception to the various changes in business practices brought about by the digital revolution. The legal profession is now more effective and efficient because of the introduction of digital processes. There are also worries, meanwhile, that digitization might have the opposite effect. These potential negative effects will be discussed in the subsequent chapters.

Law firms are rapidly implementing digital workflow automation and processes to increase productivity and efficiency. They are doing it because these technologies have many advantages over conventional approaches, such as enhanced staff collaboration and communication and more working freedom.

All these factors have come together with the demand for remote employment since the Covid-19 epidemic. Asynchronism was another legacy of the "Work from home" period. For instance, complete duties as part of a team without interfering with the teammates' work. These traits compelled law firms to embrace new work practices and emerging technologies that permitted remote and asynchronous work. The remote team can sometimes be even more productive and effective than the in-office pre-Covid working style because of legal technology.

The way lawyers work has been altered by digital workflow automation and digital procedures. By automating some duties and sending out reminders before deadlines, lawyers can concentrate on more critical work, improving the efficiency of the many processes used by law firms. Collaboration within legal firms has been simplified for lawyers due to digital methods. By writing, sharing, commenting on, and revising papers using digital technologies, attorneys can finish their deliverables more quickly.

For instance, researching a consumer before entering into a deal with them might take a lot of time if done haphazardly. In the case of a Politically Exposed Person, an automated process can make sure that the most crucial tasks are carried out, all pertinent documentation is acquired, and additional information is requested (PPE). Naturally, all of the outcomes are then documented for upcoming audits. While doing so, KPIs are being generated to identify bottlenecks and alerts are being raised when service level agreements (SLAs) are not reached.

Another example can be found in deliverables for customers. They will often go through a multi-step writing process, starting with a junior lawyer's initial draft, followed by a senior lawyer's review, then editing (for spelling, grammar, and corporate identity), before being sent to the client for approval. Additionally, new versions are frequently generated after a back-and-forth with the client. It is crucial that the technology stack used by lawyers stores every version, remark, and history.

A legal firm will probably need to take a few formal actions to close and invoice the services after they are finished. Most of the time, the lawyer wastes a lot of time on these activities to which they do not contribute. A lawyer's technology should use automated method to enable the administrative managers to accomplish various activities as soon as the lawyer advises to close and invoice.

Legal automation types

Several types of legal automation can be implemented by any firm looking to improve its efficiency:

- **Electronic signature of contracts.** A digital agreement has become commonplace thanks to the development of secure, globally recognized electronic signatures, much of which may be automated.
- **Contract automation.** Lean legal teams usually choose to implement contract automation workflows over scaling expensive manpower for low-complexity legal agreements.
- **Contract review.** Legal teams frequently use AI contract review companies to determine what is in their contracts when faced with thousands of unstructured Word or PDF contracts. In the near term, this

does fix a problem, but in the long run, it would be better to stop producing unstructured PDFs.

- **Automated billing management.** One of the main goals for many in-house legal teams is to keep spending on outside counsel under control. Numerous recent companies provide automated solutions to assist in cost monitoring and reduction.
- **Automated workflows.** To make common procedures scalable, no-code automation builders assist legal teams and law firms in fusing numerous manual processes. This could resemble the receipt and triage of new inquiries, which aids in giving legal problems the appropriate level of priority.
- **Automated risk and compliance management.** There are several ways to assess and manage a company's risk and compliance exposure, especially in highly regulated sectors like FinTech.
- **Knowledge management.** Although knowledge management (KM), used by both in-house legal teams and law firms, is a fairly mature product area, workflow automation's influence has grown, making it faster and more effective for information to reach the right people at the right time.
- **Obligation management.** For both large and growing organizations, post-signature contract administration, in particular dealing with contract renewal, can be a severe problem. This procedure can be automated to save time and lower danger.

Workflow automation and digital procedures have significant advantages in the legal industry. Technology, in particular, can significantly improve productivity and efficacy for lawyers. These tools, for instance, can speed up the process of filing paperwork, case status monitoring, and automating lawyer–client contact. Automated systems can also assist lawyers in understanding case data and making better conclusions.

Automation benefits in the legal industry

This technology increases a legal team's capacity and the caliber of their service by allowing them to concentrate on challenging issues and face-to-face interactions with clients. Automation can also refer to software robots or bots that perform basic tasks much more quickly than employees would. Bots may execute numerous tasks, like sending and receiving emails, filling out forms or scanning documents, logging into programs, creating reports, and much more. They are the ideal remedy for businesses where staff members waste time on monotonous tasks.

Data management and processing are complicated subjects, especially when laws like the GDPR are involved. Professional secrecy is another obligation for legal professionals.

You may utilize encryption and other security-enhancing technologies with a well-built automation system. Bots typically don't keep important information. Additionally, it is simple to check the steps a bot performed and find any issues because each bot leaves a trail of its activity in the system logs. Last but not least, automation provides the option of real-time process monitoring and delivers alerts as soon as a risk is detected.

However, a lack of security is not what causes most data leaks. Human error is to blame for it. Employees might misplace their access card, leave their computer in a coffee shop, use a weak password, or put their trust in the wrong person. As it makes no decisions, robotic process automation is impermeable to human mistakes and poor judgment. Robots only carry out the tasks programmed to carry out.

The main advantages of legal automation are a rise in production and a considerable reduction in time spent. Depending on what the lawyers in issue decide to do with the time they recover, this can have some clear advantages.

Lawyers may spend more time on the strategic, commercial job they were trained for without spending hours on administration and paperwork. Once a procedure is standardized and mechanized, a lawyer no longer needs to worry about it. Once automated, the identical NDA can be prepared instantaneously from a template, eliminating the need to draft it every day.

When low-value activity is automated, the cost savings might be somehow passed on to the customer (internal or external), improving their interaction with legal. Heavily manual procedures, such as wet-signature signing or manual due diligence checks, typically do not record any data, making it difficult to learn from such procedures or to incorporate them into others, like sales software. Data can be collected through automated procedures, improving workflows, and enabling analytics.

A team may be able to avoid adding more personnel by automating a process like a contract creation. This makes it possible for legal to support the firm even as it grows without its workforce. These advantages motivate an increasing number of attorneys to think about adding legal automation to their teams.

However, when automation is utilized properly, it can be quite beneficial, but it won't be able to do every work and won't advance without constant assistance from knowledgeable team members. The best approach to begin with legal automation is to start small, as with any process modifications. Massive new technological rollouts that appear out of nowhere are highly likely to fail.

It is much preferable to select a pilot program and engage with a vendor to immediately offer value for this little process change. Once you have the support of the stakeholders involved in the pilot program and the method has proven successful, you may scale it and try to transform processes at a larger scale.

Cybersecurity

Cybersecurity is becoming a requirement instead of an advantage for legal firms. Information can be kept secure, and customers' minds can be at ease when using technology to increase security. Additionally, law firms must prioritize cybersecurity when they undergo a digital transformation. Twenty-five percent of law firms indicated in the ABA's 2021 Legal Technology Survey Report that they had suffered some sort of data breach in the previous year.

Law firms must invest in cybersecurity technologies to ensure privacy compliance and the protection of all data as they face demand from customers to do more of their work online. Only about half of legal firms, according to the survey, have cybersecurity protocols in place, and some don't even have a full-time employee responsible for overseeing security.

Every law firm should have a quality cybersecurity policy to protect itself and prevent any potential cyberattack. However, estimations show that every other law firm lacks one such policy, leaving most businesses vulnerable to harmful scenarios. A successful cyberattack may result in downtime, a potential ransom payment, GDPR penalties, and, most importantly, a loss of clientele, stock value, and brand reputation.

An unexpectedly high number of GDPR breaches were discovered in the first year of implementation, according to a report published in September 2019 based on information obtained from a freedom of information request to the ICO. Out of 212 breaches, 48% of the top 150 UK companies disclosed a data breach. Even though 41% of the breaches were just the result of emailing the incorrect person, they are nevertheless considered GDPR data breaches. Although the ICO has not yet imposed a hefty fine, it is simply a matter of time.

The way other industries respond to the growing cyber danger is similar to the legal industry. Standardizing and centralizing company procedures and methods of operation, as well as enhancing the use of technology, remain top priorities for all businesses.

But law companies are particularly complex since they have numerous custom systems, bespoke apps, and case management systems that frequently provide functions that are duplicated by other systems within the same company. The threat surface expands, simple cyber hygiene practices like patching become more expensive and complicated, and the number of necessary security measures rises.

Not to mention that attackers take advantage of tried-and-true coping mechanisms, and a busy legal firm might easily ignore these hazards by coasting along in short-term survival mode. Although the structure of law firms makes them an accessible target for cyberattackers, organizational culture may not always show this awareness.

A cyberattack or data breach that results in financial loss for a law business also causes problems with operational skills. That has further hidden financial consequences by causing holes in a company's operations. One indirect method that law firms lose money following a cyber intrusion is through the loss of billable hours. These are the hours wasted in the course of a cyberattack recovery. One company in the SRA report claimed to have lost billable hours worth £150,000.

Loss of client data results in a loss of efficacy and trust, in addition to the immediate financial costs, for a law company. Compromise of customer data is not merely a breach of security; it is also a breach of trust in a field where reputation, prestige, and word of mouth are crucial to success.

The fact that law firms safeguard both client money and client data offers a special opening for online criminals. While cyberattacks can purposefully target particular people and client money, according to the SRA review, companies have also fallen victim to attacks intended to gather and control a legal firm's data.

Legal procedures must safeguard data against loss, damage, and unauthorized processing. Law firms are starting to realize that being aware of efficient cybersecurity also entails offering customer protection. Those rivals with certification are not risk-free, but they are more likely to recover if reputation is important in the legal industry. Competitors won't just need to spend less time and money; they'll also have a better chance of preserving their reputation.

It's crucial to have the ability to evaluate risks, foster an environment where they are actively anticipated, and integrate risk reduction into business practices. Understanding the many types of threats businesses face is the first step in lowering cyber risk.

Daily routines and employee awareness will be the largest area of vulnerability for most businesses when it comes to cybercrime. Most businesses said their greatest cybersecurity risk was their employees because they were easily distracted, ill-trained, and dissatisfied, which can open the door to assaults that could be detrimental to their operations.

The services provided by outside IT companies to law firms range from sporadic, ad hoc help to complete reliance. However, if law firms outsource their cybersecurity, they miss out on a crucial chance to empower employees who will likely be both the strongest line of defense and the biggest danger.

One of the most crucial capabilities a company can develop is the capacity to recognize and respond to a cyberattack at all levels. The ability to react depends on readiness: businesses that have taken steps to reduce future risks, such as controls, processes, and policies, are more productive in the event of an attack. In more than half of these cases, the cost of installing these measures was less than the initial loss the company suffered as a result of an attack.

Ways to enhance security in law firms

How can law firms defend themselves from something you're unaware of? A law firm needs to be aware of the hazards to safeguard the privacy of sensitive data against cyberattacks.

Even though the cybersecurity system is currently secure, this might not still be the case a week, month, or year from now. Never is a company "secure enough." Hackers are becoming more skilled at getting around security measures, which makes it even more important for security and awareness to develop over time.

Security awareness is essential, but for it to be effective, all personnel must receive training on the subject. Even if a designated individual is in charge of security at a legal office, it takes more than one person to keep an entire firm safe. Instead, a company's strong cybersecurity and privacy culture result from the combined efforts of all of its employees.

While many security and privacy breaches are unintentional, many of them are caused by employees' actions. Internal security breaches can be prevented very effectively by raising security awareness and promoting a security culture at work.

The firm's present security budget might not be sufficient to meet all of its cybersecurity requirements. If so, it's time to increase the budget to close this gap because failing to spend enough money could wind up costing a corporation more money.

Legal fees, compliance fines, as well as a tarnished reputation, a loss of current clients, and difficulty attracting new clients for the law firm are all possible consequences for a legal firm that experiences a data breach.

File sharing is one of the biggest threats to the cybersecurity of your law practice. A law firm's attorneys and support staff will have to share confidential material with clients, colleagues, law clerks, co-counsel, and opposing counsel during the various stages of litigation.

Because email and other consumer-grade file-sharing services like Google Drive, Dropbox, and OneDrive are not well equipped to block access to the information if it is ever stolen, these businesses cannot rely on them to share this information. What's worse still? Sharing private information and papers over courier services puts them at risk of having them lost or ending up in the wrong hands.

Legal firms require an enterprise-grade file-sharing solution. Businesses can encrypt sensitive data with the correct secure file-sharing solution so that even if it is stolen, the offender won't be able to decrypt the data they take. The private information will not be disclosed.

Firms must consider a solution's usefulness for their lawyers before selecting it. They won't employ a solution if it is too complicated. Every legal firm requires a balance between security and usability.

While most businesses could sigh with satisfaction that the tech was working from home and they could continue to provide services to clients, it was difficult to ignore how serious and vital it was to take into account the cyber risks they were facing. The level of cybersecurity needed for lawyers to safely practice from home is now present in many law offices:

- Ensure your legal firm has a clear reporting mechanism in place.
- Create strong passwords with two-factor authentication.
- Check for the security of all devices used for work purposes.
- Switch the encryption on all devices used for work purposes.

- Utilize mobile device management (MDM) to lock devices or delete data if necessary.
- Have a Virtual Private Network (VPN) to provide an additional layer of security.

Implementing cyber security into company culture

Although most cyberattacks and data breaches result from human mistakes, a law firm's strongest line of defense against cybercrime is knowledgeable, well-trained staff. Understandably, employees are afraid of making a mistake and having disastrous repercussions for their employers given the increasing number of horror stories in circulation.

To create an open, "no-blame" culture, businesses must not only actively encourage their staff to speak up about their problems and experiences but also recognize and promote such behavior. The effectiveness of cybersecurity policies will undoubtedly depend on cultivating a positive culture, but it will also play a significant role in the profession's overall performance.

Something more urgent will always require your attention. However, no law practice can continue to put off considering cybersecurity. Instead of making cybersecurity an afterthought when creating a legal firm's digital footprint, it must come first.

The company can move beyond the annual, check-the-box cybersecurity training by allowing employees to select their preferred learning style through many training techniques, including tests, quizzes, eLearning, games, videos, pdfs, and audio stories. Your cybersecurity rules will have a much greater impact if you implement brief, immersive, focused training that is little and frequent.

Employees need to be informed on current events in the field of cybersecurity. Legal companies should share information as needed and keep thorough records of ways of handling their cyber problems. Businesses frequently wrongly assume that by informing staff members about their security procedures, behavior changes. Companies should connect the dots with their staff members and be clear about their expectations.

CUSTOMER EXPERIENCE

Customers who have adopted internet channels in many other areas of their lives can provide legal firms an advantage thanks to technology. Instead of going to a law office in person, clients could opt for sending papers digitally or holding a video conference. Law firms must recognize the components that benefit their clients the most and focus their digital transformation efforts there initially.

For instance, real estate lawyers might begin by digitizing contracts and embracing online notarization. Litigation attorneys may embrace software that facilitates online depositions and remote evidence exchange. Every business has different client needs and requires unique digital solutions.

In the past, the law department did not own most of the issues and deal data required to provide insights; instead, it was spread across law firms, unstructured, and inaccessible. Today's legal departments have access to solutions like HighQ, Brightflag Workspace, M365, Evisort, Onit, Legal Tracker, SimpleLegal, and more that give them ways to access issue and transaction data, as well as precedent and previous counsel's recommendations.

Technology must continue to move data under the jurisdiction of legal departments. Companies must first gain control of their data before prioritizing data strategy and building data lakes that are central repositories for storing structured and unstructured data at scale. And the more effective it is when matched with market data, the better the internal data is.

Not in numbers, but in images, data is alive. Market leaders are concentrating on bringing data to life for their teams and consumers as a result of the availability of more comprehensive data.

What was formerly communicated through textual reports can now be done using interactive dashboards, maps, and other tools. For instance, dashboards and interactive platforms provide legal departments graphical depictions of trends across offices, business units, and competitors and instruct staff members on trade secret laws. Business leaders can see the current regulatory environment in target jurisdictions by accessing regulatory heatmaps that the legal department keeps updated.

Data-focused occupations are expanding along with data. The importance of data professionals inside the legal ecosystem is expanding along with the volume of legal data.

There are many businesses with separate legal operations departments. One of the duties of most legal operations departments is data analytics, including collecting and mining data to support the business and strategy.

Additionally, experts reveal that legal departments are spending 30% more on data scientists and adding new positions. By 2026, there will be 2–4 million data translators working in the US who act as a link between traditional data scientists and business stakeholders. These roles will bring the discipline and abilities required for better data gathering, analysis, and decision-making.

The deliverable is now completely different. The days of handwritten drafts on yellow notepads are long gone, and the same is true of legal advice that lacks market knowledge and is presented in lengthy memoranda.

Insights from legal and market data, presented with visuals and produced from the work of lawyers and data specialists, are valued by the market today. More data will be collected by legal departments, more tools will be available, and there will be more chances for the profession to change as a result of the data-driven transformation as this strategy gains traction.

Law finds it challenging to keep up with the pace of business and satisfy clients. There are too many attorneys overseeing and providing legal services, and there aren't enough senior executives with business, technology, digital transformation, and change management skills. That is a major contributing factor. The legal industry also lacks professionals in risk management, supply chain management, project and process design, customer experience, and data analysis.

Many legal service providers confuse being client-centric with caving into client demands. Both are distinct; the latter is proactive and strategic, while the former is reactive and tactical. No matter on approach, personalized services, multidisciplinary, data-backed solutions to business difficulties, and legal procedures facilitate and speed up a business.

The conversion of law to a customer-focused organization is a difficult task. It begins with cultural evolution and involves questioning the viability of its sacred cows, including its workforces, delivery structures, models, conventions, procedures, mindsets, KPIs, and reward systems. They are redesigned, improved, or replaced with new ones that better reflect and serve customers. The legal system was designed with attorneys in mind. The legal profession, of which lawyers are a part, must be developed to benefit both clients and the general public.

Law still has a culture and remnants from when it was solely about lawyers. At that time, legal knowledge was its only component, the legal profession was an industry, practicing law meant providing legal services, and the classic law firm partnership model ruled the legal world. The industry continues to evaluate performance from the standpoint of the lawyer rather than the client.

Its Holy Grail is profit-per-partner (PPP), not net promoter score (NPS). The profession's hubris, deafness to shifting customer expectations, and lip service to "innovation" are expanding the gap with its clients and providing a huge opportunity for client-centered providers to capture market share more quickly.

The growth of a small number of top legal service providers can be explained by the profession's misalignment with its clientele. The new-model providers have an incentive structure that rewards success, both internal and customer-based, deep, broad multidisciplinary experience, technological and process savvy, capital, effective use of data, and corporate DNA. These service providers have developed business-disciplined, client-focused, data-based, multidisciplinary, agile, scaled delivery of legal services in response to an unmet market need.

Customer-centric legal services

In-house legal departments, law firms, or law corporations (sometimes known as alternative legal service providers) can all provide client-focused legal services. The source(s) of the provider(s) may change, but their fundamental traits never do. These consist of the following:

1. The tradition of putting the customer's needs first
2. Client-oriented culture
3. Dedication to enhancing consumer value, happiness, and experience
4. Investment in technology, people, and training to keep up with evolving client demands and expectations
5. A method of developing long-term, non-transactional relationships with customers
6. An end-to-end supplier or a seamless supply chain used for delivering customer-centric legal services
7. An emphasis on what the client requires rather than what the provider offers
8. Success metrics that benefit customers rather than internal stakeholders
9. Results and not inputs are what is appreciated
10. Legal and compliance data integration into consumer strategy for fashion
11. A thorough data strategy that enables client-focused legal service providers to develop and become proactive business partners and value producers for their clients
12. Litigation avoidance and reducing contracts to shorten the sales cycle are two examples of customer-centric legal services
13. Data strategies for legal are not separate from business strategies, particularly for internal departments
14. Consistently excellent, hassle-free client experience
15. Work at the pace of business rather than the law
16. Senior managerial positions have legal experts not admitted to practice law
17. Digital customers need legal service suppliers who are traveling the same path
18. A capacity for mass delivery
19. Distinctive goods and services that address customers' problems
20. Dispel the misconception that law is distinct from other businesses

Customer experience management

Since the Covid-19 epidemic began in early 2020, a lot has changed, and a lot will never be the same after it is over. Customer experience management for law firms is one of the things that have altered for good. During 2020 and 2021, as more of our lives compelled online, many things were lost, yet not everything. One of the benefits is the capacity to provide clients with real, ongoing, and palpable empathy through outstanding digital experiences, and when they have experienced it, there is no turning back.

We are seeing the fourth industrial revolution, which experts have long recognized. If your company is not keeping up, it will eventually fall behind.

Keeping up with the times may be the subject of a book, not just a quick post, but a few critical components that need immediate attention and vastly enhance the client experience:

- Using a digital-first strategy
- Putting in place a website that is responsive to mobile devices
- Obtaining thorough client information to provide customized service
- Using AI technology wisely and strategically

Those businesses that are most effective at converting leads are familiar with the notion of customer journey mapping. The ability to provide virtual feedback has evolved over the past 18 months and became a vital component of this process. Timely, user-friendly virtual feedback methods are more valued than in-person alternatives by both present and potential customers, according to recent studies. Therefore, it stands to reason that businesses that offer these services attract this clientele.

A modernized grasp of customer experience mapping is just the tip of the iceberg. Reflecting modern ways in the client research tools decreases the time and money typically spent on stale techniques. Customer experience management must now be a lean, focused division of your business, just like your legal services, and it cannot be bloated or wandering.

Customer experience management is not something that law firms are educated in. Your lack of experience becomes a bigger liability since the success of law firms depends more and more on customer-centered business strategies. An expert legal consultant can not only assist businesses in changing their strategy but can also make sure that they focus their time and resources on offering the highest quality legal services.

ARTIFICIAL INTELLIGENCE

Technology can provide lawyers access to information beyond what they can learn from endless documents. Solutions are available that "read" papers using natural language processing, finding errors, contradictions, and even emotions or forgeries. The software can sift through data to look for patterns that could provide significant knowledge and insights. Advanced, AI-driven software can also assist lawyers in developing a case or restructuring a contract (Figure 11.2).

Artificial intelligence (AI) advancements are opening up fresh opportunities that will increase technology's usefulness in the legal industry and possibly lower the cost of getting legal assistance. Now that artificial intelligence has been developed expressly for the legal sector, it may be able to accomplish activities that previously required human intelligence. It offers the possibility of a sector transformation, which could alter the lawyer's function.

The application of artificial intelligence (AI) by lawyers and within the legal sector is only now starting to take off. What effects will this technology

Figure 11.2 **AI and text processing.**

have on the legal industry? Artificial intelligence adoption, especially by in-house lawyers, will drive a transformation in the practice of law during the next few years.

As email revolutionized daily business, AI will spread like wildfire and become every lawyer's necessary assistant. Those who resist the change and do not adopt it will fall behind. Those who succeed will eventually discover that they have more time for the two activities of thinking and giving advice, for which there never seems to be enough time.

The simple explanation for the steep rise in AI spending is as follows: Humans may focus on tasks that bring value, those that computers really cannot do or do effectively, by dismissing them from regular tasks that computers can perform, resulting in significant productivity gains and cost savings.

AI enables the search for concepts, such as contract review and analysis for due diligence, to detect changes in email communication tone. These include searching for code words used to conceal the true nature of the conversation and even mapping out where the computer comprehends what needs to be drawn and prepares the adequate document.

So far, the legal sector has benefited from artificial intelligence in several ways, including data processing, legal research, content generation, and reduced stress. Artificial intelligence has also increased productivity in the legal sector, providing attorneys more time to concentrate on their cases and clients rather than boring paperwork.

The underlying expectation is that their legal department will follow suit as C-Suite executives become more intelligent and involved in AI usage for business operations. The potential to use AI in the legal sector and specifically in legal departments is poised to move machines beyond simple keyword search tools and transform them into partners with whom lawyers will collaborate to provide the business with better, quicker, and less expensive legal services.

Artificial intelligence in legal industry

There is no doubt – artificial intelligence is changing the legal sector. However, its primary function is to take over and supplement the clerical labor of legal professionals so they may focus on higher-level activities, including contract negotiations, court appearances, client outreach, and more. When pairing up artificial intelligence with machine learning (ML), legal firms can take their digital transformation process to the next level.

The effectiveness of document analysis in the legal sector improves the usage of AI-powered tools. At the same time, machines can evaluate papers and declare them to be pertinent to a specific case. ML algorithms accomplish their work of locating more papers that are equally significant after certain categories of documents are flagged by machines as relevant. Machines, which are better at sorting through papers than humans, lighten the weight of examining all the documents on human workers by presenting only the dubious documents.

Legal research is tedious and very time-consuming. Research thoroughly and on time is crucial to winning a lawsuit. Thanks to technology, AI systems may now do legal research by using natural language processing to assist in document analysis.

Every law practice in the world employs legal support personnel to carry out due diligence and gather background data on behalf of their customers. To provide their clients with appropriate assistance, attorneys must do due diligence, which includes verifying facts and statistics and carefully evaluating judgments rendered in earlier instances. Due diligence is a lengthy task for humans. With the help of AI tools, these legal assistance specialists could complete it much more quickly and accurately.

A substantial amount of the work done by law firms is contract review. Law companies edit contracts, point out errors, and advise clients on whether they should sign the agreement. They can also help clients negotiate better terms.

The largest law firms can study and review contracts both in bulk and individually due to AI. This technology has helped law companies in sifting contracts more quickly and accurately than human lawyers. More accurately than human attorneys, AI and ML with data analysis capabilities can predict the outcomes or verdicts of judicial procedures. Legal counsel will be able to respond to these queries with confidence and in a far better way now that AI is in place and has access to years' worth of trial data.

A divorce and its settlement is a drawn-out procedure that could take a year or longer and possibly cost thousands of dollars, depending on the chosen attorney. The traditional divorce process has evolved into a courteous approach that offers a nearly self-guided divorce solution for a price that is one-fourth of what lawyers quote, thanks to the advent of AI and ML in the legal sector.

Couples can explain their ideal outcomes via the AI-enabled online method. The AI-powered system would guide them through several modules, and all of the crucial choices are conceivable given the circumstances specific to the couple. Additionally, you can get assistance from legal professionals who are prepared to intervene when necessary and provide their advice.

Data processing and research in the legal profession have been enhanced by artificial intelligence. It can aid in the lengthy and difficult process of discovery, which is one of the most time-consuming parts of legal work. By searching through data and information to find pertinent papers, AI can facilitate this process.

Since artificial intelligence is designed to work with patterns and vast amounts of data, it can enhance data processing. Artificial intelligence can:

- Examine unstructured data from case-related information,
- Determine when rules change and alert legal teams when terms change as well,
- Look up key terms in papers and other languages that are pertinent to the case,
- Identify inconsistencies in contracts that can cause issues later in the legal process.

Artificial intelligence (AI) cannot take the role of humans in legal research. However, it can improve them by streamlining the most time-consuming, data-related components. It enhances the research capacity of legal teams by employing algorithms to sort a large volume of data.

Although AI assumes time-consuming jobs, it won't replace attorneys. Instead, it will give them more time to concentrate on other activities, enabling them to focus on developing their cases and interpersonal relationships with clients.

The legal industry is known for requiring a lot of time, and lawyers frequently experience stress and sadness, which can severely impact their performance. By minimizing the long hours and stress that come with the job, artificial intelligence might greatly boost morale.

Some people worry that jobs may be replaced by faster, more reliable technology as a result of artificial intelligence taking over professions. In many industries, such as the legal one, it is not the case.

For legal teams and attorneys, artificial intelligence has considerable advantages because it reduces the time needed for data processing and legal research while boosting morale and time management. Additionally, by putting less effort into monotonous chores, attorneys can focus more on developing their cases and the more intimate components of the legal process.

DIGITAL TRANSFORMATION BENEFITS IN THE LEGAL INDUSTRY

Law firms today confront many business issues, such as increasing competition, client demands for higher value, pressure to increase profitability, slowing revenue growth, and cybersecurity risks that could damage their and their client's reputation.

Many legal firms and in-house counsel believe that digital technologies like cloud computing and artificial intelligence (AI) hold the key to innovation, better utilizing new market opportunities, boosting business productivity, and lowering risk.

There is no requirement for user-owned hardware with cloud computing. It is replaced by a cloud service that requires a monthly subscription but offers cost savings and better protection for the company and its customers. Additionally, it facilitates seamless file exchange and collaboration between attorneys working on the same issue, doing away with the need to send endless reams of paper back and forth between offices.

Just consider your benefits from increased productivity brought on by automating monotonous chores. You would be given the power to have immediate insights into open issues, complete team contributor visibility, and the information you need to make quick decisions. Counsel is then in a proactive position to advise clients since they are confident that nothing unexpected is going to happen.

The recommended way to future-proof your business and better respond to customer service disruptions, remote working circumstances, problems in our reviving markets, and improved productivity and time management is through digital transformation, which applies to in-house counsel and legal firms. Admit that having more free time is appealing, regardless of whether your goal is to draw in new customers or strike a healthy balance between work and leisure.

Every organization that has successfully implemented change is aware that a well-thought-out plan and employee buy-in to the shift are essential components of success. Companies can benefit from the significant operational and financial advantages and boost employee satisfaction by putting into practice a digital transformation strategy.

Simple digitalization, like automating corporate processes or establishing a paperless workplace, is only the initial step. Complete organizational digitization is labor-intensive and time-consuming, but it may lead to numerous improvements, including upgrading your business models and procedures, mining data for insights to serve customers better, and transforming employee perceptions and connections with technology.

To guarantee enhanced productivity, client data protection, a consistent working environment, and a positive user experience with this type of platform, modern, fully integrated, and secure platforms are crucial for digital transformation. On the other hand, legacy systems rely on the disjointed process of patching several software applications throughout the company to form a single business environment.

The likelihood of a successful digital transition increases when internal users adopt "digital as their default." Modern computer platforms are easy to use, safe, effective, and productive, all contributing to a great user experience. Comfort and ease of use for purposes like online shopping need to be replicated by these technologies.

Make sure there are sufficient reserves of resources for innovation (e.g., for technologies such as AI). The business will be able to concentrate on activities that help it gain a deeper understanding of customer needs and highlight new ways to use technology and organizational skills to meet customer requests as quickly and effectively as possible due to this planning.

Since cybercriminals view law firms as lucrative targets, preventing security concerns, breaches, and data theft should be a top priority. Utilizing concepts like need-to-know security, policy-based controls, strong encryption, and records management is essential for operationalizing security. Data protection will become more commonplace and inconspicuous by implementing security by design to generate various defenses. Currently, AI and analytics offer the best ways to automate detection in a way that is both affordable and prevents security breaches.

Streamline operations

Businesses should look to technology to eliminate pointless business processes and streamline operations to remain competitive in cutthroat areas like the legal one. Companies utilize Business Process Management (BPM) software and solutions to automate data gathering, analysis, and report generation for putting the best procedures in place. By optimizing operations, new technology removes pointless duties so that one may concentrate on crucial work.

Casework consolidation

Digital transformation solutions are necessary due to the changing nature of litigation. Since the legal sector was among the first to adopt solutions like e-Discovery in the workplace, it gave the impression that it was technologically savvy. Modern digital tools are necessary to improve online information access, nevertheless, as the sector becomes increasingly digitized.

Technologies like these can be used to condense casework and enable e-Discovery, making it simpler to manage access to information.

Automate tasks

Companies can use cloud technologies, artificial intelligence (AI), and digitally-based technology to automate the processing of paperwork rather than doing it by hand. Automation has the major advantage of reducing human intervention in corporate processes, which can lead to human mistakes. Automation eliminates the need for disaster recovery and saves time and costs as a result.

Resource management

Information and resources are consolidated as a result of digital transformation, giving legal practitioners better control and accessibility. Many digital technologies can combine software, databases, and applications into a single entry point for efficient resource management. In the legal sector, keeping track of resources and paperwork is crucial for project planning and execution. Digital transformation in modern technologies makes resource management simple.

Excellent customer experience

Every business strategy should be built on the client experience. The chances of a firm succeeding without happy clients are small. Client interaction is a crucial component of the legal profession, so clients should have a wonderful experience. Clients can communicate freely with their legal counsel and feel confident that their personal information is secure by utilizing secure solutions that encourage client participation and security.

Additionally, offering outstanding customer service and the means for customers to discuss their experiences (such as forums and online reviews) motivates additional potential customers to look into your company.

LIMITATIONS OF DIGITAL TECHNOLOGY IN THE LEGAL INDUSTRY

Both individuals and employers must acquire new technology and capabilities due to digital transformation. Despite being crucial for success, digital transformation can be difficult for non-technical staff to understand and use. User adoption is a major obstacle in the digital transformation process.

In addition to looking into user-friendly technologies, employers must offer training to staff members. It's common for businesses to adopt new technology in the belief that it would enhance x, y, and z, without considering the possibility that users may not be able to utilize it entirely.

If the people involved are not prepared, disruptions in how any profession is practiced could result in significant changes. Similar to other fields, the legal industry will not fully benefit from technological improvements if practitioners do not accept these changes and reinvent their abilities.

Worrying about what this technology might do to the industry is another aspect that contributes to the lack of urgency. The potential for data breaches, which could jeopardize client confidence and result in potential legal issues, is one cause for concern. Additionally, not all legal firms are eager to invest in upgrading their existing IT infrastructure. Given the uncertainty around the return on investment, they might be reluctant to invest resources.

However, relief comes with carefully reviewing the subsequent actions. After all, if fear continually wins, change will never occur. Additionally, the culture that enforces lawyers' linear thinking and propensity for traditionalism should be dismantled. They must keep up with the demands of updating their business models and presenting fresh prospects for creating value.

Businesses must transition to digital. However, not all of them have the tools and abilities necessary to do so. As customer expectations shift toward instantaneous services and digital processes, they will soon expect their legal counsels to follow suit. Law companies must gradually transform themselves to be more client-focused and technologically advanced to succeed in the future.

Being a fully digital attorney could also be harmful. If you work as a fully digital professional, several factors can kill your productivity. Digital distractions, such as checking social media and asking "what happened on Twitter in the last 15 seconds," are one of them (Instagram, TikTok, Facebook, Snapchat, and the list keeps growing). It is easy to lose 30 minutes when taking a quick break by checking several apps.

Finding information across various sources is another productivity killer. Did I, for instance, receive the material via email? A WhatsApp message, perhaps? When you require this information to fulfill a deadline, having an ad-hoc flow from several sources and storage locations is a headache.

In a nutshell, a legal firm's ability to operate effectively and efficiently depends on the digitization of its procedures. But not in any way at all. Disorderly digitization causes issues distinct from manual paper-based labor but just as critical.

Cloud migration late adoption

Managing client data is undoubtedly a delicate subject for law firms and other organizations in the professional services industry. Law firms regrettably embrace most modern technologies slowly, so most of them still use traditional technologies for their billing, CRMs, data management, and security systems. That can make moving to the cloud a challenging and expensive procedure.

Fortunately, there is a solution, and we can implement a unique strategy for your cloud architecture that best suits your company's requirements, scope, and budget.

Online security

One of the major worries for law firms is security, particularly with the rise in remote operations. The fundamental problem here is that while most reputable companies are well-prepared for security challenges inside the workplace, very few can make the same claim regarding the state of remote or hybrid work models. The cloud will also provide the answer because it has the opportunity for the highest levels of security. To lessen the security hazard, a move to the cloud must occur concurrently with adequate personnel training.

Staff training

New skills, continuing personnel certification, and training are necessary for cloud adoption. That may seem like a big hurdle for law firms that prefer to stick to their core competencies and avoid dealing with IT modernization. Businesses must realize, nevertheless, that adopting technology solutions will free them to pursue prospects for future expansion and cost-cutting initiatives.

Current solutions and future integrations

Although they acknowledge the necessity for modernized integrations in the future, organizations frequently insist on preserving functional outdated systems. For instance, a business might have used the software provided for hosting emails, managing projects, and other virtual meetings. Now, it must incorporate each of them into a single solution while still allowing for future tool additions.

CONCLUSION

Digital transformation is a corporate change in the legal sector. Businesses in the legal sector must implement user-friendly digital technology to stay profitable and competitive. Although the process has significant challenges, such as learning and adaptation, the outcome is worth it.

Although the digital transformation process is ongoing, the long-term effects on business growth make it worthwhile. It makes sense that when digital innovation is adopted, lawyers should acquire new skills and create effective delivery methods.

Many technological tools can assist in document drafting, legal research, document disclosure in court, legal counseling, and online conflict resolution. Instead of performing more regular or ineffective duties, technology can help lawyers spend more time practicing law. When examining the best ways to increase the sector's efficiency, it's not just the lawyers who need to be considered. Additionally, we should consider improving its usability for clients and individuals who need legal counsel but are unsure how to receive it.

Digital tools will never be able to fully replace the human capacity for lateral thinking and the resolution of complex legal issues. Instead of seeing digitization as a threat to their existence, lawyers should view technology as a way of enhancing their toolkits and enabling them to do their duties even more effectively.

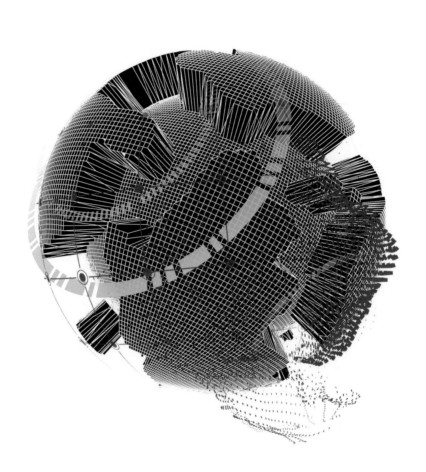

Chapter 12

Digital transformation in the insurance industry

Digital transformation is one of the buzzwords in the entire business world. Numerous commercial sectors have been altered by it, including manufacturing, retail, hospitality, and healthcare. Although the insurance sector has historically been sluggish to modernize, it is now more apparent than ever that insurers should embrace digital transformation.

In order to meet client's requests, insurers have had to digitize many parts of their businessess. Any company that wants to remain competitive in today's market must provide for customers when and where they need it. Digital transformation of insurance enables insurers to do just that and will continue to transform the market for years. It is driven by artificial intelligence (AI), machine learning, mobile service, predictive analytics, live chat, etc.

There are innumerable instances of how the insurance sector has evolved from digital transformation. Operations are streamlined, consumer interactions occur over chat, claims may be handled automatically and brokers can collect all their data to do their tasks more quickly and accurately.

For the insurance sector, these innovations are merely the beginning. Insurance companies must use digital transformation technologies and develop innovative ways to optimize their operations to be ready for a digital future. Ninety percent of insurance executives say they have a well-thought-out, long-term plan in place for technological innovation.

Regardless of the goods or services, it is considerably more accessible for customers and prospective customers to switch from their existing supplier. Today, many applications and websites make it simple to compare prices for goods and services online, obtain advice from friends and family, and ultimately post user reviews of goods or services. To do all of this, they typically use smartphones.

Businesses today must modify and alter their current business strategy if they want to grow or maintain their consumer base because this exponential transformation is so extreme. How does this work? This is accomplished by demonstrating a commitment to innovation and being willing to upgrade and alter their outmoded, 20th-century business model. The typical business model for insurance companies entails creating and distributing packages through a network of agents.

DOI: 10.1201/9781003305163-12

The industry must create a corporate business strategy that places digital technology in the correct perspective to support that strategy if it hopes to reap long-term benefits. This is important since digital transformation entails ongoing innovation. Contrary to popular belief, technology does not come first. Technology is merely a tool used by a corporation to accomplish its objectives (Figure 12.1).

Since communication is pivotal in digital organizations, insurance companies must be able to connect their portals and systems. Insurance businesses have already deployed and used different portal types: partner, enterprise portals, and intranet. Modern customers interact with insurance businesses through a variety of channels. All channels must be combined to form an omnichannel digital ecosystem.

Insurers must combine all the elements that can increase consumer involvement at different levels. Through omnichannel apps, users search, recommend, and share. Self-service web pages and a review of their performance, policies, pricing, maturity, and payment dates are helpful for clients

Figure 12.1 Insurance industry and AI.

and partners. For insurance businesses, having all the necessary information in one location is becoming essential.

There are limitless opportunities for insurance businesses that reconfirm their dedication to innovation and successfully deploy creative solutions. They must collaborate with fresh partners to use an innovative environment to adopt new paradigms. This will make it possible for businesses to stay current with trends and lead the market.

The insurance sector has a long history of being infamously slow to adopt new technologies. It once appeared inconceivable to think about streamlining processes, switching to an online user interface, or even processing claims on the same day; that is, until the global digital transformation of all businesses, which has now included the insurance industry.

"The process of employing digital technologies to develop new – or adapt current – processes, cultures, and customer experiences to meet shifting business and market requirements" is how insurers define digital transformation. Digital transformation is this reinvention of business for the digital age.

Property and casualty insurers frequently do the majority of their business online or through digital applications today. Simply put, consumers today want speed. They want instant gratification, and when it comes to insurance that means quick policy quotations, quick bill payments, and quick communication.

The future directions of the insurance sector's digital transformation have yet to be discovered. But according to experts, the Internet of Things (IoT) and new AI developments will significantly affect claim processing and underwriting. These tools and others are a part of the current insurance industry's digital transformation process.

REASONS TO IMPLEMENT DIGITAL TRANSFORMATION SOLUTIONS

Digital transformation of numerous business sectors is now a reality thanks to the IoT. The insurance sector, one of the most established and slow to absorb new trends, has been redefined by digital transformation.

Consumers today generally expect quick service, whether for food delivery or handling their insurance claims. More and more consumers are completing claim forms on their smart devices, uploading supporting materials, setting reminders, getting updates, and making insurance policy payments.

As a result, digital tools are becoming more prevalent to support insurers' customer communications and speed up the insurance process in general, including policy quotations, payments, and paperwork. Early adopters are already reaping the rewards of digital transformation in a few areas of the insurance sector.

Driving excellence and efficiencies

Insurance companies must strive for cost savings and long-term advantages whenever possible in an environment of low interest and growth. That is made simpler by introducing new strategies, including integration, APIs, automation, system changes, outsourcing, shared services, and partnerships in digital transformation. Moving operations to the cloud and implementing dynamic resource and asset utilization are two different approaches to changing insurance organizations.

In light of this, digital transformation powered by AI is probably applied in digitized claims processing, improving customer satisfaction.

Personalizing service offerings

Digital transformation has been molded by customers' wants and desires to encompass a more individualized service range and product variety. The previous "one size fits all" paradigm is no longer relevant because of add-ons and customized insurance solutions. Customers can receive tailored packages specifically catered to their demands thanks to AI.

Master technology disruption

Businesses in the InsuTtech sector are leaders in developing novel business models. New firms are now entering the market, equipped with tools like robotic process automation, advanced analytics, and AI. By establishing synchrony between diverse processes, digital transformation boosts revenue, productivity, and sales.

DIGITAL TRANSFORMATION TRENDS IN THE INSURANCE INDUSTRY

Insurance businesses used IT in the past to reduce expenses. The focus right now is on how technology can assist companies in accelerating growth and enhancing consumer engagement. Trends influencing the insurance industry's future are developing as more insurance businesses transfer their operations online.

Insurance has evolved into a digital innovation hub because of FinTech investments and InsurTech companies. To improve the experiences of customers, employees, partners, and other stakeholders, insurers must embrace change and rethink their business models to transition to a compliant, secure, and digitally connected operating model. The development of new products, digital experiences, and the transformation of crucial processes, such as marketing, distribution, underwriting, claims, finance, and accounting, are all simplified by new digital tools and capabilities.

DIGITAL TRANSFORMATION USE CASES IN INSURANCE

Insurance customers' demands are changing. Insurers must be prepared to adjust to shifting work habits and turn to technology that may make interacting with digital insurance firms more comfortable and easy for everyone.

By 2025, AI and machine learning will have automated 25% of insurance activities, according to Mckinsey. The insurance sector is a suitable candidate for digital transformation since it is choked with manual procedures and inefficiencies in areas like customer service, fraud detection, underwriting, claims processing, and policy administration.

Underwriting is one of the fundamental steps in the insurance process, in which the insurer optimally values and diversifies the risk. Compared to humans, AI/ML models are faster and more accurate in interpreting historical data. As a result, insurers may streamline underwriting and increase profitability by applying these models.

To evaluate the risk of prospective harm, insurers acquire data on their insured customers or real estate. For instance, underwriters use the following variables to determine risk in the case of vehicle insurance:

- Value of the insured vehicle,
- Driver's license history,
- driving preferences of the insured,
- the insured's age and driving history,
- The area's crime rate.

Data interpretation is also necessary for the diversification strategy. It could be problematic for insurance companies to insure comparable assets in the same area. Think about a company exclusively offering house insurance in regions with frequent, destructive earthquakes. Due to the claim increase, such a corporation may experience financial hardship following the earthquake. Because of this, insurance companies require large databases to diversify their plans appropriately.

Claim processing involves several stages: review, inquiry, adjustment, remittance, and claim denial. Insurers must handle several papers for each layer, which may be done automatically with document automation systems. Insurance companies may use document automation to automatically extract data from documents, spot fraudulent claims, and confirm claims that adhere to the policy.

Based on business regulations, insurers utilize text analytics to identify fraudulent claims by feeding predictive analytics systems with data from the claimant's narrative.

An efficient strategy to improve customer experience involves designing a self-service portal that customers and insurers can visit to obtain answers to problems, perform business (i.e., transactions, orders, filing a claim, paying bills, etc.), check on status, submit support tickets, and download materials.

DIGITAL TRANSFORMATION SUCCESS STORIES IN THE INSURANCE INDUSTRY

The level of competition is rising. Market newcomers include tech titans and insurers focusing only on the digital space. Customers are raising their expectations of insurers, who must deliver a better customer experience at the same time.

Insurance companies must change if they want to compete. They must put greater emphasis on pleasing customers and become more customer-centric. Additionally, they must adopt new technology that will make them more quick and effective.

One of the most successful examples is Zurich Insurance Company, which decided to maximize the potential of digital transformation, particularly AI, with the help of a conversational AI company. The result was the intelligent virtual assistant named Zuri, tasked to address all client inquiries by providing 24/7 availability of customer care.

This intelligent virtual assistant allowed the insurance company to permit consumers to manage current policies quickly, facilitate quicker inquiry resolution by directing clients through each phase, and handle regular chores rapidly, such as submitting a claim or withdrawal, updating personal information, requesting a callback, etc. In terms of results, the digital transformation solution automized 85% of all online communication, with 70% not requiring human assistance.

Another excellent example is an insurance company from the Netherlands, which followed a B2C approach, aiming to invert the current players in the market. As part of its first international expansion, Lemonade introduced its ground-breaking policy, nicknamed Policy 2.0 for its potential for disruption, in Germany.

Lemonade is a different business from traditional insurance providers since powered by digital experiences, bots, and AI. Consumers in Germany may now obtain liability and content insurance quickly, whenever they want, file claims promptly, and receive payment instantly through any channel.

After all, although most conventional insurers are still cautious, customers are no longer ready to put up with poor service. With one-click buying, immediate customer support, real-time updates, and on-demand access, they are accustomed to simplicity and speed. How else to compete with giants like Amazon, Lemonade, and Insurify?

A financial services company, Nationwide, partnered with Insurify, which provides comparisons of home, life, and auto insurance, to power its digital brokerage division.

Customers can use Nationwide's brokerage platform and Insurify's comparison tool to get a policy from a different carrier when they need coverage that Nationwide cannot supply. Despite not being the carrier, Nationwide will continue cultivating connections with customers by assisting them in finding coverage and providing a satisfying customer experience.

Nationwide's investments in innovative InsurTechs pave the way for future joint solutions. Nationwide aggressively supported new companies and collaborated with them to build its digital capabilities. By developing these investor ties, Nationwide may supplement its services with InsurTechs' solutions, as was the case with Insurify, in which Nationwide made its first investment in 2017.

Metromile, a pay-per-mile digital insurance disruptor, hopes to revolutionize with its recent integration of Dwolla's digital payments platform. Metromile is now able to empower insurers all over the world to provide a more streamlined and client-driven claims experience. These include Real-time Payments via the RTP® Network, Automated Clearing House (ACH) transfers via the ACH Network, or Push-to-Debit disbursements, by integrating Dwolla's payment technology into its proprietary no-code claims automation platform Metromile STREAMLINE.

Claims processed using this system will provide a superior client experience in terms of automation and convenience but will also expedite the process, offer new levels of claims accuracy, boost staff productivity, and lower loss adjustment costs.

Customers frequently have to wait an extra week to access their funds and deal with transaction costs nearly ten times higher than those related to digital payments because 75% of claims are paid using old-fashioned paper checks. Ask a GenZer what a check is to understand what a throwback this payment method is today.

To better meet the demands of a changing economy and accommodate new travel behaviors, Metromile is also innovating its product line. Due to the significant increase in courier employment following the COVID-19 outbreak and the increased use of ridesharing services like Uber, insurers must modify their product lineup to accommodate these more erratic work schedules.

To this aim, Buckle, a business that specializes in meeting the insurance requirements of delivery and ridesharing drivers, has teamed with Metromile. With the help of REPORT, a digital, contextual, self-service FNOL solution from Metromile for consumer-facing data gathering and customer care agents, Buckle can now provide its clients 24/7 access to vital services and assistance for digital claims via mobile devices or the internet.

Customers' demands of their insurance companies are changing. Brands in this market need to be prepared to adjust to shifting work habits and turn to technology that may make interacting with digital insurance firms more comfortable and easy for everyone.

BENEFITS OF DIGITAL TRANSFORMATION FOR INSURERS

Different sectors are slowly but surely being affected by the digital revolution. Businesses across all industries are adopting cutting-edge technology to reinvent their operations. When it comes to willingly challenge the existing quo, the insurance industry is one of the most robust industries. What are the requirements for the insurers to approve digital transformation?

They must study, among other things, what benefits it may provide the insurance sector. No matter how large or small an insurance organization is, every part may be transformed and digitalized. A few benefits of adopting digital transformation include the following:

- Giving live chat and mobile support.
- Building data pipelines to expedite predictive analytics.
- Increasing fraud detection.

Improved efficiency

There are hundreds of procedures that make up the insurance industry. There are at least one or more repetitive tasks in each process. The most common source of mistakes, which frequently result in project setbacks and delays, is repetitive duties. The main goal of digital transformation is to streamline and improve all processes.

How does digital transformation increase the efficiency of various insurance business areas? It involves more than just locating a software application or platform. Actions aiming at reinventing the procedures and best practices now in use are included in digital transformation.

With machine learning's capacity to learn and generate appropriate insurance plans, imagine never having to redraft a single insurance policy. Consider utilizing a platform that automatically handles and settles disputes. What if an advanced chatbot took straightforward client queries while referring irate customers to human agents?

The only project that can assist in speeding up both front-end and back-end operations right now is digital transformation. It goes beyond just implementing new technology and aids insurers in implementing new procedures with their objectives and clients' requirements in mind.

Personalized service

Customers nowadays differ from those of a few decades ago. They understand the value of ease and would happily put their faith in a business that makes it simple for them to use the services. Personalization, one of the critical components of the customer-centric strategy, is added on top of it.

Tailoring services to a single customer's requirements and expectations are becoming standard practice for a competitive insurance product.

There is, however, one issue here. Numerous insurers have tens of thousands or even hundreds of thousands of customers. A tailored experience can only be provided by using all the available resources. With the aid of digital transformation, insurers can tackle this problem head-on and allocate resources more wisely.

Modern platforms allow brokers and consumers to utilize a single system for all transactions. Customers may submit payments for the invoices online and get immediate responses to their claims. Brokers also have a comprehensive picture of a customer, historical claims, and agreed policies.

Systems using AI and machine learning can gain knowledge through experience. These technologies can personalize the offers and automate delivery because every consumer interaction and transaction is recorded.

As a result of data analytics, marketing teams in the insurance sector may use insights to adapt offers to specific consumer categories, increasing lead generation and conversion rates.

Building data pipelines

Data pipeline construction is the foundation of digital transformation. This phrase has undoubtedly come up throughout your investigation. It does raise the question of why the data pipeline is so important. There is a massive quantity of data produced by the insurance sector. It originates from internal systems, clients, and marketing teams and frequently from wearables that employees and clients sport. Maintaining the current data-collecting methods results in what experts refer to as data silos.

Large volumes of data are kept in "data silos" that are exclusively accessible to specific departments and frequently only to high-ranking employees. This behavior has grave consequences because:

- Having the same data saved in different locations can waste essential resources.
- A limited perspective of the data prevents access to an enterprise-wide picture.
- Inconsistencies frequently result from poor data integrity when data is stored across many databases.
- Collaboration is impossible when departments cannot communicate data, making it impossible to support cooperation.

A single data pipeline is built by insurers with the aid of digital transformation. A data pipeline is what? Various data processing operations are in a data pipeline. It can be set up to suit the requirements of the particular insurer. Insurance companies can gather, process, and store all data on a single platform thanks to a data pipeline. It alludes to the information in a pipeline. Any number of things, like your quarterly financial reports, broker success rates, client encounters, and feedback from IoT data, is possible.

Insurance companies may utilize the applications to acquire immediate insights and strengthen their business choices thanks to the processed data being in one spot. Additionally, it makes room for predictive analytics, which may be used to find trends and spot profitable business prospects.

Improved risk management strategy

Like any other sector, insurance operates in a dangerous environment. A variety of technical tools can assist insurers in managing very particular risks. A data-driven risk management plan may be developed with the help of digital transformation on all fronts.

Insurance companies can gain a market advantage by reducing the risk of fraud. The sector has been plagued by fraud for a while now. Only now, insurers have access to technologies that would have assisted them in shifting from a reactive to a proactive approach.

Planning for development and expansion may be made easier for insurers by embracing digital transformation. Insurers may now use data pipelines to obtain real-time insights rather than relying on information from third parties, a hunch, or experience when making choices. The data loss failsafe systems can also aid insurers in avoiding business interruptions, which frequently incur costs they cannot afford.

LIMITATIONS OF DIGITAL TRANSFORMATION IN THE INSURANCE INDUSTRY

Although the insurance sector was somewhat late to adopt digitalization, following COVID-19, the industry began to experience disruption from the advent of digital technology. More than merely paper form digitization and operational optimization are now being discussed.

Instead, the rise of new business models, income streams, and enhanced consumer experiences is being driven by digital transformation. Innovative technologies and digital ecosystems that expand insurers' access to new markets and create novel business prospects are driving this transition.

However, these businesses still need to work on articulating how their new digital strategy fits into their longer-term objectives.

You will always need to integrate the current procedures into the increased capabilities if your business is a startup. Your outdated systems cannot just disappear. The risk of losing clients exists otherwise. Additionally, because technology is constantly developing, it is crucial to prepare a new unit for changes in the future. Additionally, unforeseen challenges like the worldwide pandemic are frequent in our lives.

Having your reaction team or a trustworthy insurance software development provider who will handle your company's continual transformation is the winning way to ease this suffering. Although your customers might be eager to begin their improved digital adventure on your new platform, they may still need to move to it. Because of this, it is essential to provide simple interfaces that can be customized for use with many platforms, systems, and gadgets. Your staff should receive still more training if you want them to function well in the modern insurance environment.

For insurers, digital solutions also bring new norms and exponential data growth. Take catastrophe modeling as an illustration. An insurance provider will need information on houses and companies that might be in danger to manage risks. In the future, this will result in publications covering a more extensive range of hazards.

The explosion of data and variables will also result in higher criteria for data quality and unified interpretation. You risk getting unreliable results and forecasts if there are discrepancies in the variations. Consider a single claim number that may be given to an automobile accident or each participant involved.

With two main attack types present in this context—vulnerabilities in the target systems and data theft or manipulation—it isn't easy to undervalue the significance of solid security for all phases of the insurance process. You never know when your website may stop working or whether there will be a security breach involving storing the customers' data.

As a result, insurance firms are mainly accountable for protecting their data systems. On the path to digitally enhanced insurance, the issue of few resources might be a significant roadblock. Where can businesses obtain additional funding to invest in innovations while fighting to survive in quickly shifting markets?

Yet, it is inevitable that new companies will enter the market the very next day if insurers do not accelerate the pace of global digitalization trends.

Digital transformation challenges for insurers to overcome

Any company's capacity to satisfy the demands of its clients by offering excellent products or services is the foundation of its success. Today, however, insurance companies use digital platforms, such as websites and mobile applications, to serve their consumers, which is becoming increasingly important. This development is compelling insurers to develop and carry out digital transformation strategies to satisfy the market's changing demands.

However, there are many obstacles that companies must get beyond to implement a digital transformation strategy effectively.

Maximizing the value of customer data

The capacity of insurance firms to properly gather, handle, and keep client data presents their first barrier. According to market research, consumers are now prepared to trade more personal information for higher advantages like money or specialized services. According to Accenture polls, 50% of insurance consumers are open to sharing more personal information in exchange for better premiums, cheaper interest rates, or other financial advantages.

Insurance companies gather clients' personal information but encounter barriers that make it challenging to evaluate this information to provide the sort of individualized service that clients need and to operate more shrewdly. Overall, sophisticated AI systems are required to process consumer data efficiently. The insurance industry's inability to automate back-end procedures that would enable AI systems to interpret user data has made these reforms challenging.

Regulations governing the collection and use of personal data present another challenge for insurance businesses. For instance, the General Data Protection Regulation (GDPR) influences how all European enterprises gather, use, and keep personal data. The use of the gathering of personal data may become more challenging and demanding as a result of GDPR's strict regulations for businesses.

New competition

The emergence of new competition in the form of online banks and insurers is the second issue confronting insurance firms. Today, a sizable portion of the population is prepared to buy insurance from these neo-banks and neo-insurers.

Over 60% of insurance clients are willing to purchase policies from neo-insurers or other IT firms just entering the market. When merely focusing on Millennials, who will continue to make up a more comprehensive section of the client base for insurers, the percentage of consumers wanting to move substantially rises to over 80%.

These neo-insurers are appealing because of the 24/7 availability of a wide range of services, goods, and information online, especially for younger customers like Millennials and others who are comfortable utilizing digital services.

It is crucial for established participants in the insurance sector to give their consumers access to a full range of services, products, and information online to counter the development of these neo-insurers. Second, to set themselves apart from neo-banks and neo-insurers, conventional insurers must provide consumers with other services and advise them in their lives.

The same survey estimates that 48% of banking consumers want their banks to act as their advisors when making big decisions like buying a home, taking out a loan, or getting a new automobile.

Of them, 76% of insurance consumers said they would want pertinent information from insurers that would enable them or senior family members, for instance, to live longer and safer lives in their own homes.

Adopting innovative technology to improve the customer experience

Adopting or implementing cutting-edge technology that can enhance the client experience is another issue that insurance businesses must overcome.

Insurers must embrace advanced technology, such as intelligent sensors and other IoT gadgets. IoTs allow insurance firms to gather crucial consumer information to safeguard or determine their risk levels and give them more individualized service.

Incorporating AI technologies like chatbots or virtual assistants will be essential for insurance firms' ability to serve consumers' demands. According to Accenture, 74% of insurance clients can get help from chatbots or virtual assistants whenever needed. Despite the advantages, the sectors still need to improve by using AI technologies like chatbots or automating procedures.

Cybersecurity

Cybersecurity is a great difficulty that insurance firms confront. Insurance businesses must always be on the lookout for threats due to the nature of the data gathering and the vast volume of it to safely and ethically protect consumer data and their online platforms.

Cybersecurity concerns significantly influence customers' choice of insurance provider. Nearly 31% of consumers in the insurance sector claim that data protection plays a significant role in determining how confident they feel about a brand's cybersecurity. Accenture reports that 43% of clients attribute most of their insurance loyalty to data protection.

CONCLUSION

For insurers, becoming digital is a need, not a choice. Insurance companies must embrace technology and utilize it to their advantage to compete in this new era. The process of technological change is not linear. While some organizations could change their procedures completely, others might introduce new tools gradually.

Insurers must need help with offering digital services and retaining a personal touch. The secret is maintaining alignment with your organization's goal and core principles while upgrading operational models to promote optimal effectiveness and customer happiness.

Customers today need best-in-class digital experiences to be the standard, which is why businesses across all industries are accelerating their digital transformation. Because there are so many alternatives on the digital market, brand loyalty has reduced, and patience has waned in the age of immediate gratification.

To meet the expectations of insureds, insurers today have had to digitize many parts of their business. Any company that wishes to remain competitive in the market today must provide for clients what and when they are in need. Insurance is undergoing a digital revolution, enabling insurers to do that and will continue to reshape the market for years. This transition is fueled by AI, machine learning, predictive analytics, mobile service, live chat, etc.

For the insurance sector, these innovations are only the beginning. Insurance companies must use digital transformation technologies and innovative methods to improve their operations to be ready for a digital future. Ninety percent of insurance executives say they have a well-thought-out, long-term plan in place for technological innovation.

Chapter 13

Web 3.0

WEB 3.0 IN GENERAL

Web 3.0 and digital transformation go hand in hand. New technological components like Web 3.0, NFTs, and the Metaverse have significantly impacted how companies today approach digital transformation. Web 3.0 has made the virtual world as close to customers as possible. Not only is this technological leap helping people interact with each other in the virtual world – it is helping them promote their products and services, forge new connections, and expand business opportunities. Web 3.0 is the third generation of the World Wide Web, and it is marked by a shift from static web pages to a more interactive and dynamic web experience. This shift is made possible by advances in web technology, such as AJAX and HTML5, which allow for more sophisticated web applications.

DOI: 10.1201/9781003305163-13

Unlike previous generations of the web, which companies like Google and Facebook centrally controlled, Web 3.0 applications are decentralized and distributed. This allows for a more democratic and open internet, where users have more control over their data. Today, Web 3.0 is being used to power a variety of next-generation applications, such as virtual reality (VR), augmented reality (AR), and artificial intelligence (AI).

Web 3.0 has fundamentally changed the internet as we know it today by enriching it with several new characteristics. Today, we can say that Web 3.0 is:

- Verifiable
- Trustless
- Self-governing
- Permissionless
- Distributed and robust
- Stateful

Important to say is that Web 3.0 has native built-in payments, whereas Bitcoin is often referred to as the internet's native currency.

The most significant novelty when approaching Web 3.0 from a development standpoint is that applications are not built and deployed on a single server and do not store their data in a single database. Web 3.0 applications, often called DApps, run on blockchains, decentralized networks of numerous P2P nodes (servers), or a combination of the previous two.

Even though networks are decentralized, they are incredibly stable and secure. This is mainly due to network participants being incentivized to provide services of the best quality to anyone using the said service. This point is where cryptocurrency enters the entire story. Cryptocurrency acts as a financial incentive to users who participate in creating, governing, or working on a specific project.

Also, cryptocurrency plays a vital role in many Web 3.0 protocols. In essence, Web 3.0 stands for a set of protocols led by blockchain where the logic of the internet is combined with the logic of the computer. This enables users and developers to change how the internet is connected in the backend.

There are numerous protocols in Web 3.0 that offer a wide array of services, ranging from computing and storage to bandwidth, hosting, identity, and other services. Important to note is that in Web 2.0, cloud providers often provide such services. Web 3.0 aims to change this and enable users to truly live and evolve with the internet.

Every protocol has its own set of participation rules on the technical and non-technical levels. What drastically differs from the Web 2.0 landscape is that users can earn significant amounts of money by simply participating in the protocols.

To summarize, Web 3.0 technologies are changing how businesses operate and communicate with customers. In addition, they are also helping to drive digital transformation of industries ranging from retail to healthcare. Web 3.0 is still in its early days, but it is already impacting how we live and work in the 21st century.

How Web 3.0 is revolutionizing payments

With Web 3.0 payment infrastructure decentralized, users have gained control over the technology provided by the said decentralization, not by a company or a government. The third iteration of the World Wide Web provides greater data security and eliminates the need for centralized intermediaries.

This is the most significant t difference between Web 2.0 and Web 3.0 payment infrastructure. Web 2.0 payment infrastructure heavily relies on centralized banking systems that include virtually everyone with an open bank account but exclude all those who don't want to open a bank account or simply do not have access to the centralized banking system due to any reason.

Web 3.0 payment infrastructure welcomes all users to participate financially without needing a centralized intermediary. Payment providers worldwide see Web 3.0 as an opportunity to tap into a newly emerged source of money flowing across the internet. One of them is Paypal which, according to its 2022 reports, aims to allow its users to trade cryptocurrencies via the Paypal platform.

In essence, Web 3.0 payment infrastructure eliminates the "middleman," thus guarding the users' privacy and allowing true democracy to flourish on the internet. Using the Web 3.0 payment infrastructure, users can send money to any user they want – anonymously, swiftly, and securely. While big tech took advantage of Web 2.0 by creating their payment processors and developing a complex payment infrastructure, Web 3.0 gives power back to the user.

What many internet users have wished for years is a highly minified bureaucracy. Until recently, users had to register for financial services with a centralized intermediary of their choice to send money to another person (or any other entity). By doing this, users have enabled these intermediaries to access their transaction data, placing anonymity and security on the sidelines. In Web 3.0, all data is encrypted, ensuring the security of users' transaction data.

How Web 3.0 payments work

To participate in the Web 3.0 payment infrastructure, users must own a Web 3.0 wallet, often called a crypto wallet. Web 3.0 wallets store fungible tokens and NFTs (non-fungible tokens), enabling users to interact with DApps across numerous blockchains.

Almost every Web 3.0 wallet today can be used as a browser wallet, a browser extension, or a mobile wallet (a DApp built for iOS and Android devices.) Whatever digital asset users want to store, they will need a Web 3.0 wallet to do so.

Every Web 3.0 wallet comes with a set of two keys, a private and a public key. The public key represents the wallet's address and is used to send and transfer tokens from wallet to wallet. Public keys are safe to share with others; sharing them will not endanger the user's wallet security. On the other hand, the private key is used as a password to perform any transaction in the wallet. It should not be shared, as it can easily endanger the security of the Web 3.0 wallet.

Even though the Web 3.0 wallet comes in several different forms, the browser extension is the most popular Web 3.0 wallet type. Numerous trading platforms, OpenSea most notably, support browser wallets, and users find them the easiest to use.

To summarize, Web 3.0 payments differ from Web 2.0 payments in three key areas:

1. Unlike the traditional payment systems where processing fees are high, Web 3.0 payments are stunningly fast, resistant to censorship, and provide enticing rewards to users.
2. Web 3.0 payment system is immune to financial censorship. It offers a decentralized and permissionless protocol that allows developers to build DApps and continuously improve. Everyone can participate in the Web 3.0 payment system.
3. There is no need for third-party involvement. Web 3.0 payment system garners trust, enabling users to communicate privately and publicly, but it happens without involving a third party. All this happens thanks to smart contracts, which self-execute when all set conditions of that smart contract are met.

In essence, the Web 3.0 payment system raises the bar and opens new doors for companies trying to provide better service. Implementing it, however, is a long and challenging process, requiring both time and expertise in the latest Web 3.0 technologies. Even though it may seem like a tiresome process, it's a huge selling point. Whenever you have a chance to improve your service by implementing new technologies, you should consider it. This is one of the key thoughts you should carry after reading this chapter.

Impact of decentralized autonomous organizations on work

Decentralized autonomous organizations, or DAO for short, are one of the crucial novelties of Web 3.0 and have become a vital piece of the blockchain ecosystem. In essence, they are organizations without centralized leadership. DAOs function as horizontal structures, thereby being entirely governed by community members. A set of automated rules must be built into the protocol for DAO to function without needing leadership.

Unlike traditional organizations, where the most critical decisions are made at the top, DAO emphasizes community governance. This makes DAO fit for the social impact sector, and a considerable number of non-profit organizations are starting to see the value of DAOs due to this fact. Until now, the community-based decision-making process was easily overlooked, but the emergence of DAOs created an entirely new business environment where every vote counts.

Through their use of smart contracts and governance protocols, they have enabled numerous communities to take ownership of causes that are of great importance to them. Here is an example. A charity organization usually has a Board and a certain number of members. But when deciding how to spend the available funds and which activities to organize, the decision is often left to the Board. Decentralizing this charity organization creates new opportunities for the community. In such a decentralized organization, every member has a direct say in how the organization is governed.

Of course, with such new concepts, there are always several drawbacks to pay attention to. For instance, including too many people in a single decision raises fear of the organization being slowed down. Also, in DAOs, there is always a risk of a 51% attack, which means that 51% of the votes are being harmfully manipulated to take control of that DAO. Despite those fears, many DAOs exist today. They make their decisions efficiently, continuously improve, and end their business years with profit.

The importance of smart contracts

When writing about Web 3.0, it's imperative to delve into smart contracts. Smart contracts are computer programs hosted and executed on the blockchain, enabling users to complete any activity, such as payment, without the involvement of third party. The code in smart contracts defines the conditions that trigger a particular outcome when the conditions are met. When running on a blockchain, smart contracts enable multiple parties to reach a shared result securely, swiftly, and accurately.

They are integral to Web 3.0, especially DAOs, which could not exist without smart contracts. Important to note is that smart contracts are highly beneficial regarding automation, as there is no administrator to control all processes. The code continues to run without interruption as long as it is hosted on the blockchain.

How do smart contracts work

In many cases, centralized single-source-of-truth systems are more vulnerable to attacks than decentralized systems, with smart contracts on the blockchain eliminating the risk of an attack. Additionally, smart contracts can increase efficiency, provide process transparency, and minimize operating costs when applied to a multi-party agreement.

The importance of smart contracts, as well as their purpose, can be seen rather clearly when we take a look at how smart contracts are built. When creating a smart contract, first, it is required to define the business logic. Before programmers even start writing the code, they need to work with business experts to define the criteria for the desired behavior in case of accurately defined circumstances. Both developers and business teams need to encode complex operations using sophisticated logic; where an example would be if they want to encode determining the value of a derivative financial instrument.

When developers start writing smart contracts, they need to use a smart contract writing platform to write the contract and test its logic. In larger organizations, separate teams are dedicated to testing smart contracts and ensuring top-level security. In smaller organizations, the most common occurrence is developers testing their smart contracts. In some cases, an organization may even employ a smart contract expert whose sole purpose would be vetting smart contracts and ensuring their resilience to external attacks.

After the team tests the smart contract, they can deploy it to an existing blockchain or distributed ledger infrastructure. A key piece to this puzzle is an oracle, a data source constantly streaming data between off-chain databases and smart contracts that are "on-chain." This breaks an old myth that blockchain is ineffective as all data must be located on the blockchain.

Unsurprisingly, there are usually more oracles connected to a single, smart contract. This is common in the blockchain landscape, as we can imagine oracles as simple data streams where each oracle helps bring in different sets and types of data. To trigger smart contracts, certain events need to happen, and oracles help ensure that smart contracts get executed by providing the necessary data. Once the smart contract obtains the necessary combination of events from any number of oracles, the smart contract executes.

This shows how vital smart contracts are for the development of global business. They allow for no influence on the decision, regardless of who or why is trying to influence it. They are immune to the failure of large sections of the network they run on and are resilient against malicious hacking. Smart contracts are a unique phenomenon in our lifetimes, and their value in the world's future can be easily recognized.

What exactly is a blockchain?

After we have gone through several key aspects of Web 3.0, it is beneficial to double down on the most recognizable term, the blockchain. We can define blockchain as a distributed database, or a ledger shared among a computer network's nodes. In essence, a blockchain is a database that stores information in a digital format. The data structure is the most notable difference between a blockchain and a traditional database.

Traditional databases structure data into tables, while blockchain structures data into blocks, hence the name. These data blocks have specific storage capacities; when they are filled, they get closed and linked to the previous block. This forms a blockchain. All new information that comes in is added to a new block which will, in turn, fill up and connect to the previous block, thus creating a longer blockchain.

Crucial to note is that when a certain block is filled, it's set in stone. Data cannot be altered once it is written down in a block of data, and it allows us insight into the timeline of data. We know precisely when a specific data set was added to the blockchain, further empowering much-needed transparency.

Lastly, the goal of blockchain is to allow the recording and distribution of digital information, but without the possibility of editing that information. This ensures complete transparency, especially when dealing with payments.

The transaction process on a blockchain is straightforward. Once a new transaction is entered, it's transmitted to a network of peer-to-peer computers scattered across the globe. This computer network solves various mathematical equations to confirm the transaction's validity. Once the transaction is legitimate, it's clustered together with other confirmed legitimate transactions in blocks. These blocks are then chained together and form a permanent transaction history. When all this happens, the transaction is finally complete.

Decentralization

In previous chapters, we had a chance to review the benefits and drawbacks of centralized and decentralized systems. In the following example, we will go deeper into the decentralized nature of blockchain.

Imagine that a particular company owns a server farm housing 1,000 computers for maintaining a crucial database with essential data. This server farm is located in a building also owned by the company. Even though it is possible to think of this as a good solution, it provides a single point of failure. For example, the data gets lost or corrupted if the electricity goes out, the internet connection breaks, or a fire engulfs the building.

Blockchain fixes the issue of a single point of failure due to its decentralized nature. It allows data in that database to spread among network nodes at different locations. This maintains the fidelity of the data stored among network nodes. For example, if someone tries to change one data record at only one instance, the other nodes remain unharmed and prevent attackers from altering data records. All other nodes cross-reference each other and pinpoint the node tampered with. This way, we have all the information on when and where someone tried to alter the data.

This mandates that data such as transaction history are unchangeable and irreversible. Once written on a blockchain, it stays forever, immune to external influences.

How secure is blockchain?

The shortest answer would be that blockchain is highly secure. Blockchain achieves security in several ways. In this chapter, we will go through every single one. Firstly, new blocks are created and lined up chronologically, meaning that a blockchain always has an end. After we add a new block to the blockchain, it is nearly impossible to change the data in one of the previous blocks. This is possible only when 51% of the network agrees to change the data.

Furthermore, let's imagine a lousy actor is running a node on a blockchain to steal cryptocurrency from other users. When this hacker tries to alter their copy, they will soon hit a brick wall. When all other users who run nodes on that blockchain network cross-reference their copies, they will immediately notice that the hacker's copy stands out and that the hacker is attempting something malicious. In this case, the hacker's copy of the blockchain is cast away.

The hacker must control most of the blockchain copies to achieve such a feat, at least 51%. These attacks are unlikely to happen as they require vast resources due to all the blocks that need to be redone with new timestamps and hash codes. Taking into account the fact that there is a huge number of cryptocurrency networks that are growing and increasing their transaction speed, such attacks are almost unthinkable.

Let's imagine that hackers try to accomplish this malicious task. This is a huge event, and such events are never unnoticed. Many members of that blockchain network would recognize this as a malicious attack and would hard fork off to a new version of the chain unaffected by hackers' attacks. When network members decide to hard fork off, the attacked version of the token gets worthless, leaving the attackers with less than cents in their reward for completing the attack. It is easy to say that participating in the blockchain network as a member is more financially fruitful than attacking a network and attempting to steal other users' cryptocurrencies.

Successfully grasping the immense security of blockchain and its key aspects, such as decentralization and transparency, is crucial to understanding the contents of Chapter 14.

CREDIT SCORING INDUSTRY DIGITAL TRANSFORMATION

Credit scoring, also named the credit reporting industry, is one of the biggest industries in the US, worth 14.2 billion dollars. By now, it is common knowledge that it is plagued with numerous problems. This fact is highly troubling, as many people and businesses largely depend on their credit scores when asking for a personal or business loan.

Credit reporting is one of the critical parts of maintaining a good credit history. Unfortunately, credit reports often contain errors, and millions of people and businesses in the US are on the receiving end of these issues. A vast number of inconsistencies on credit reports, paired with the low percentage of people and businesses who report errors found on their credit reports, created a strong distrust in the credit scoring industry. The fact that credit bureaus that continue to make mistakes in their credit reports are the ones who create credit scores is unsettling for a large number of US citizens.

A US-based tech company sought to break that cycle of false business credit reports and low business credit scores by leveraging Web 3.0 technologies. They have outlined several critical problems they aim to solve with their technological solution; there is no clear way to obtain a business credit report, the process of obtaining a business credit report is difficult to navigate, and generated reports are hard to understand. This issue leaves many companies in the dark, where they simply don't know they have a credit report.

To solve these credit reporting issues, the company launched a blockchain platform that allows businesses to create credit reports without involving a third party. Their solution is complex, and it solves all issues of the credit reporting industry. This goes as far as rendering credit bureaus unnecessary, but many systems need to be in place to reach such an achievement.

A blockchain platform to end the monopoly of credit bureaus

The platform is a network of facts secured by cryptography and blockchain technology. It allows users to record borrowing, lending, and buying and selling activities and initiate and rate B2B relationships. The end goal of the platform is to eliminate the need for credit bureaus and enable businesses to build their business credit through everyday business activity.

The entire system works as a network focused on a decentralized reputation. Participants are pseudo-anonymous, self-sovereign entities that want to establish a reputation. To clarify, reputation refers to, in this context, commercial relationships, and transaction success. Transaction success refers to lending, borrowing, customer satisfaction, client satisfaction, and product satisfaction. The architecture consists of a web-based user interface, smart contracts on a blockchain, relayer nodes paid in the platform's cryptocurrency, a web server, and an API.

Participants are known by a blockchain address, such as an Ethereum address. The entire system is created to facilitate intentional de-anonymization for business entities who want to publicly display their company data and inject the formal analysis reports made by credit bureaus.

With many everyday business activities being unnoticed and unreported by credit bureaus, this platform offered a solution beneficial to all B2B companies in the US. The platform has a viral structure, incentivizing users to seek out other users willing to attest, on-chain, that a commercial relationship exists between them. This works as a win–win situation for both parties, as they can consider this an opportunity to build their business reputation by rating the other party.

The entire network is focused on collecting raw data and financial concerns. To limit the constraints on the network, compute-intensive operations, interfaces with external systems and verification of the identities of accounts who want to de-mask are off-chain, considered both external concerns and revenue opportunities. Rating agencies who want to verify identities and issue their credit and reputation scores can freely design their in-house processes based on their interpretation of data collected on the network.

All business entities on the platform are enticed to reveal their legal and basic company information. Still, they are not required to do so if they want to stay anonymous. Unmasking is encouraged because it will enrich reports for all business entities who do business with them. The reports instantly become more readable when a business user is identified with a legal name rather than an Ethereum address.

The data structure is where this platform takes off. It is a network graph of nodes representing identities, joins representing relationships, and histories of those relationships. The data structure is designed following the principle of minimalism, resulting in a simple and flexible structure that accommodates variation. Identities are generated in the browser and can be transferred to another browser by using the 12-word mnemonic.

Each join has several stages, each describing the lifecycle of the B2B relationship on the platform. More accurately, there are two mandatory stages and three optional stages. The mandatory stages refer to the sender requesting the consent of the receiver to open a business relationship and the receiver agreeing that a relationship exists. Going further, three optional stages refer to both parties declaring the end of their relationship, the sender rating the receiver, and the receiver rating the sender.

Immutability strongly persists on the platform. The existence of bilateral joins is immutable, as well as the fact that a proposal for a relationship can be issued and deleted if not accepted by the other party. A proposal can be deleted, but the fact that it was issued persists.

Behind the complex mathematical equations and the technological solution created is the need to create more opportunities for small- and medium-sized businesses to build their credit through everyday business activity, create their error-free credit reports, and get a realistic business credit score based on immutable data.

In more than two-thirds of credit reports issued in the US, there is an error, and often there are more than a few errors to be found. Small business owners only sometimes find out that they even have a business credit report; if they do, they do not always take care of all errors on the report. Businesses create inarguable proof of their business relationships and transactions using the aforementioned blockchain platform to record their transactions and rate their business partnerships.

This mandates that there is no user surveillance, the privacy of all users is respected, and users are in total control of their business credit reports. The platform prevents the data from being aggregated and exploited. There can be no error in users' credit reports as there is no way to alter their transaction data, as the data is located in the blocks on the blockchain. By joining and using this credit reporting blockchain platform, companies can take charge of their business reputation and credit scores since they will have complete insight into their transactions and business relationships.

BLOCKCHAIN OF THE FUTURE

A Canada-based company managed to create a breakthrough in blockchain technology, aiming to create an open, scalable, stunningly fast, and secure platform for DApps. Most successful blockchain models rely on either raw computing power (Proof of Work, Bitcoin, for example) or exogenous currency (Proof of Stake, Ethereum, for example) to commit network nodes to honest consensus. Still, the growing reinvestment cycle and economies of scale facilitated the centralization of node ownership.

When talking about Proof of Work, this resulted in enormous mining companies investing tens of millions of dollars into the infrastructure, rendering small operators inessential to the network. Before Ethereum's Merge (transition from Proof of Work to Proof of Stake) in 2022, large mining companies made mining for small operators challenging. Bitcoin mining is also primarily attributed to large mining companies due to low mining incentives after the last halving.

In Proof of Stake, the store of value transfers directly from an origin currency into the staked blockchain without needing physical infrastructure. This facilitates centralization, which in itself poses a problem. The company decided to build an alternative blockchain network, resistant to centralization and with uncapped scaling potential. They have confirmed their assumptions that their network drastically reduces operating costs, finalization time, and attack vulnerabilities.

Decentralized networks have the property of the distributed trust, which means that network can be trusted to execute all transactions honestly, regardless of network members being dishonest. This property alone eliminates the need for a third-party trust provider, like a bank. Globally known networks such as Bitcoin and Ethereum have demonstrated massive demand for this service model, but they have met their limitations.

Blockchain networks operate under a parallel computing model, which brings into balance trade-offs between three core attributes of blockchain networks; security and integrity, intended decentralization and trust distribution, and overall scalability. Usually, this mandated that networks be designed in a "choose two" trade-off, which results in one of these three core attributes not being fully evolved. Vitalik Buterin, the creator of the Ethereum blockchain, expressed this idea; thus, it has become known as the Buterin Trilemma.

Crucial to mention is that the Proof of Work model shows to be intrinsically inefficient due to its high energy consumption, which is required to continuously solve complex mathematical equations and complete the block. The central Bank of Canada found in their studies that operating the Canadian economy on Bitcoin would require 26.3 times the country's power consumption. Currently, the Bitcoin network itself consumes more electrical power than Sweden. Even though this ratio may improve, it is wise to look elsewhere for alternative solutions.

On the other hand, Proof of Stake systems displaces physical investment with economic investment. Even though this seems more efficient than the Proof of Work, it substitutes electrical power for derivative economic power.

In theory, economic power follows available energy input into that economy. As demand for blockchain operating on Proof of Stake rises, its utility and economic value require proportional energy to maintain security properties. The problem here is twofold, with energy being the first severe issue in this case and rising costs being the second. Large mining operations and pools have consolidated network ownership, facilitating centralization in open blockchain networks. This eliminates the unique property of the distributed trust and drastically reduces the network's value and security.

Proof of Stake networks is bound to be centralized in time without the physical constraint of building mining infrastructure. For open blockchains to be recognized and accepted globally as an economical medium, node ownership must be widely distributed, even in the face of centralization efforts.

To provide sustainable distributed trust, the company ensured its blockchain network resolves the Buterin Trilemma. This means that network membership is available to any node of computing power, the entire network has enough computing power to process transactions in parallel, and any malicious attacker must harness computational power greater than the entire network's computing power.

The company needed to improve traditional blockchain technologies in three areas to reach such an achievement. Regarding the architecture, the company decided to use real-time optimized order transactions block-lattice as the core data model. Blockchains are, in their nature, serial structures, while block-lattice is parallel. Adopting the block-lattice data model allowed for the parallel processing of transactions and smart contracts, massively increasing the network's efficiency.

Regarding the consensus, the company opted for the leaf consensus mechanism. The network uses the protocol to offload bandwidth requirements from nodes to clients and reduce communications complexity at the nodes. In turn, this enabled tens of thousands of peer nodes, thus garnering decentralization and strengthening the network's security. Lastly, regarding asynchronicity, the company uses its stem asynchronous clear/settle paradigm to decouple the clearing and settlement mechanisms, enabling limitless scaling.

This blockchain platform is set to revolutionize peer-to-peer payments due to its incredibly fast transactions, executed on a fully parallel network with unlimited scaling. The platform enables users to initiate and execute transactions in various manners, such as tap-to-pay, phone-to-phone, card-to-phone, code-to-phone, and internet direct. In their efforts to evaluate the network's efficiency, the company claims it is 20,000 times more efficient than the Ethereum network, currently the world's leading blockchain network. In a single minute, the platform can finalize 29 million transactions.

By solving the Buterin Trilemma, this blockchain platform easily outperforms all other blockchain platforms, steadily maintaining its decentralized property as it grows. It finalizes globally in under 2 seconds and at a fraction of the costs of other blockchains, like Ethereum.

IOT AND WEB 3.0

A US-based company ventured to create the world's first blockchain platform with a pet-focused economy and ecosystem. Their goal is to create a future where every pet is identifiable on a blockchain, and pet owners can access the necessary information from data points across numerous pet products, services, and healthcare industries. They use blockchain to securely deliver resources to rescues and shelters in need, to improve the quality of care pets receive, and to strengthen the value of the global pet community.

They have identified several key issues the pet community faces and devised blockchain technology solutions. The problems they recognized are too many kill shelters, giant corporate monopolies, pets going missing, deceptive marketing on pet-related products, and fragmented pet data.

The company's blockchain platform facilitates transparent donations delivered directly to shelters and rescues in need, thus decreasing euthanization. To battle giant monopolies such as Mars and Nestle Purina, they provide their community with the power to rate pet health-focused businesses through crowd-verification. Regarding fragmented pet data, they aim to create a normalized pet data language and ensure all collected data is compatible with the existing data, allowing the data to be more accurately interpreted.

Furthermore, they force companies to market and label their pet-related products truthfully. The platform does this by providing users with high-end analytic tools and informing them if a company they regularly purchase pet products from has mislabeled or misrepresented one of their products or services.

The most intriguing aspect of this pet-oriented blockchain platform is its physical product, the Pet Tag. The tag helps pets find their way home, but its function goes way beyond that. It updates pet information in real time and posts bounties and rewards for lost pets. If an owner loses a dog, the entire community will be aware that the dog is missing and will be rewarded by the platform if they help find the missing pet. Such a solution has yet to be seen in this market and is one of the unique blockchain platforms that leverage IoT devices coupled with Web 3.0 technologies.

The company built an entirely new and advanced ecosystem emphasizing transparency, traceability, and crowdsourced verification to create valuable insights globally. The platform's uniqueness stems from its unusual mix of physical devices, digital assets, and a utility token. Moreover, the company is revolutionizing data ownership since the platform can reward pet owners for sharing information about their pets. Pet-related information, from vet visits to pet stores and daycares, is of high value in the pet industry, and vendors and researchers are willing to compensate for such information. This, in turn, creates a circular economy that comes off as a novelty in the pet industry.

The company created an NFT marketplace to host only pet-related NFTs and enable users to mint their own pet-related NFTs on the platform. The main selling point of their NFT marketplace is its charitable nature. A certain percentage of revenue made from every NFT minted is donated to animal rescues and pet shelters, which furthers the company's goals to reduce the number of kill shelters. Besides the charitable nature of the platform, the company intends to allow users to use their pet-related NFTs in the metaverse, which is due to come in the future as the platform continues to grow.

Furthermore, crowd verification, one of the core aspects of the platform, is used as an instrument for promoting high-quality products and helping pet owners make the right decisions when buying pet products. The company's IoT devices and analytics tools are connected to the blockchain platform and relay information on pet products, trying to establish if the products are healthy and safe to use. Once a vendor gets verified, this leaves a significant impact, as the verification is possible only through positive reviews and performance. Unlike the ongoing situation on Twitter, where users can purchase their verification badge for a fixed monthly amount, this blockchain platform allows only crowd-verified vendors to gain their verified status. Besides vendor companies getting exposure to their target market, this furthers one of the company's primary goals of bettering pet health and safety.

To empower researchers, manufacturers, marketers, and veterinarians with detailed pet data, the company, created a marketplace to trade data between pet owners and interested parties. In today's world, where many call data the most significant asset a company can have, the ability to earn cryptocurrency by providing veterinarians with pet data makes a difference in how users conceive the entire platform. This manifests as a win–win–win situation for the company behind the platform, for pet owners monetizing their pet data, and for researchers and veterinarians who obtain detailed pet data that is much needed to further their academic and business efforts.

To top it all and create another revenue stream, the company launched an e-commerce marketplace offering pet products and services and company merchandise. As it is the nature of all blockchain projects to have their cryptocurrency and use it as a payment method across their platforms, the situation on this particular blockchain platform is the same. The company created its cryptocurrency token, which they use to facilitate trade between all parties on data and product marketplaces.

The company uses the token to enable the trade of products on services on their marketplaces and to entice and engage users by rewarding them for providing data and using the platform's features. A small transaction fee on all trades is implemented, and the company is amassing it and reinvesting into research and community development. This is standard for blockchain transactions, with countless blockchain platforms using transaction fees as one of their revenue sources which fuel growth and enable companies to massively reinvest into new assets.

This blockchain project specifically sets new standards for upcoming ventures. The team behind the project succeeded in creating a self-sustainable blockchain platform that helps users gain access to valuable data or monetizes their data and makes a significant dent in the pet industry. The platform's humanitarian nature is noticeable in every feature it offers, and it stands for a precious cause.

The most eye-catching aspect of this blockchain project is its wide array of components and technologies, particularly IoT devices. Unlike many other blockchain projects that do not use physical devices to collect data, the team behind this pet-focused platform ensured that their IoT device, the pet tag, can enable its buyers to receive cryptocurrency in exchange for collecting the data through the device. Even though there are already many IoT devices created specifically for pets, this company's IoT device is simple enough for non-technical users, and its monetizing nature is its most significant selling point.

This project opens a new chapter of technology and raises the level at which we integrate technology into our daily lives. Web 3.0 technologies have yet to be recognized and adopted on a large scale, but that is entirely natural, considering that Web 3.0 is relatively new and that many bright minds worldwide are yet to become familiar with all the tools and technologies Web 3.0 offers.

In the early days of the initial Web 3.0 craze, blockchain platforms with no real-world utility were succeeding mainly because they were among the first ones to venture into the unexplored territory of technology. Today, the crypto market is entirely different, especially during the bear market, where even high-quality blockchain projects with immense real-world utility are not guaranteed to succeed. Investors recognize the future potential of Web 3.0 but will not readily offer their resources to platforms they do not believe will make a significant impact.

This blockchain project is one of the most successful projects in the pet industry due to the value it provides to all its users. As with Web 2.0 and its early days, it must come to a turning point at which a global majority will recognize Web 3.0, its value, and the opportunities it creates for new industries to emerge and new technological breakthroughs to develop. Luckily, with blockchain projects like this, Web 3.0 is on a solid path toward becoming globally recognized.

Chapter 14

Next steps and what the future will bring

Instead of conclusion

Digital transformation today is not something you choose to do, but the main thing if you'd like to stay on the top of the contemporary business game. IT equipment affordable prices in last two decades allowed wider range of users. Companies are connecting networks, using laptops and mobile phones, syncing tablets…but is that enough? "New software" is producing more tasks in order to make it work and people, in the end, are frustrated with digital technology which supposed to free some time and enable them to deal with new leads, projects, or ideas.

There's no magical wand which would make digital transformation easy. Strategic holistic approach where all actors within business process are included can be a common denominator in transformation process but from there – everything else is simply – a challenge. There are some cases where challenge is too complex, unapproachable, or just not in an optimal timing.

However, that doesn't mean that one should quit, but learn from it and allow a new attempt with more focused, prepared, and clear steps which will be more likely to generate a success. Numerous people which are to be included in digital transformation is a key challenge, but also the most beautiful thing in these implementations.

It is easy to conclude that people are the first stage of digital transformation and maybe the main point which will decide if process is on velvet. However, that fact is occasionally forgotten and could lead to a potential disaster. Doing business changes for bottom up is a good thing – digital transformation should go in all possible directions, if we want to make it happen. Once the members of the management discuss and point out main benefits of digital transformation and people realize it as an advantage and not a threat, then we can talk about minimizing resistance to change. Eliminating such resistance is utopia, but sometimes it drives you even further and forces you to adjust transformation direction in an unknown or unplanned zone.

It is also noticeable that people aren't ready to give up on known processes. Digital transformation, if orchestrated right, should destroy previous processes. Three papers with same data, coming and going via email are still

DOI: 10.1201/9781003305163-14

the same process, only via IT. One input place with automatic submission confirmation is a process transformation.

On the other hand, companies, and people in charge of different sectors may be familiar with current process. They are experts in it. Unfortunately, they cannot see outside that "box" and such attitude can make transformation even more challenging. Consultants, IT experts, data managers, or analysts are support for managing people. If they are unclear in current process directions, it is almost impossible to have successful transformation.

Clear and structured strategy should involve all stakeholders. In this global, modern, and data-driven business world, it is not possible to talk about digital transformation without changing business logic. In last years, it was often seen that companies got new fancy equipment, set Wi-Fi, and couple of servers where some application is hosted. Since no process is touched and remained the same – apart from taking paper from one office to another instead of sending emails – we cannot talk about full digital transformation but only digitalization.

Changing a process is nothing but an easy task. Since people like habits, changing something they used to do is always a base for resistance. Silent and delayed resistance is most common since people aren't willing to confront directly especially in large rigid enterprises. Education and bottom-up approach are just minimizing resistance approaches – not eliminating them. Only possible thing and most commonly used, which will engage full involvement of employees is education.

By offering a quality preparation, involvement during the process and support after the implementation is done, management can assure as good as possible integration process.

Infrastructure is maybe one of the first things managers think about when talking about business software. Where the data will be stored and which technical specifications are satisfying in the long run are just some of the questions which are to be answered. Last decade is in cloud for sure but some clients are willing to invest more for start and have all data "under eye."

CALCULATION AND COMPANY GROWING PLAN PLAY SIGNIFICANT RULE IN INFRASTRUCTURE DECISION STEPS

Networking part is something which can be easily forgotten. Coronavirus showed us bottlenecks when even medium-sized company is not flexible enough to provide work from home environment within a day. VPN, firewall, user permissions, and security guidelines are not something set only for corporations. Systematic approach with planning and documentation is advised when talking about IT networks and whole ICT in general. Each router, server, and even PC or laptop must have own identity card where predictive and unplanned maintenance or repairs will be noted. It will help you decide whether to buy a new equipment or fix it for the third time in a row.

No matter on your employee number, high level of security awareness must be implemented. More than 80% of data leaches, security breaches or virus attacks are caused only by people's fault. Nothing strange if storing password on the piece of paper placed on the top of working table is a common habit. One of the greatest approaches is setting your password to "incorrect" and putting on victory smile when entering anything in the password field provides "your password is incorrect" reminder.

There's no need for being paranoid when it comes to IT security, but trying to stick to best practices is something worth repeating to all users no matter on education or experience level, gender, or company position. It is probable that your company will not be a "person of interest" and directly attacked, but even a small security issue on unmaintained WordPress website can provide you inconsiderable consequences for the business.

Selection of IT tool is another challenging point. Using a product which is not a fit for our company, digitalization goal, or field of work can do more harm than not using it at all. Selecting out of the box commercial software is a good thing if you are having standardized workflow. However, most of the companies are having their specific needs or processes which are not like other firms (maybe that's one of the reasons why they are successful). Deciding whether to choose off the shelf solution or to tailor your own would depend on most factors where money is not the only one.

Software as a service solution can be affordable, easily scalable, and simple to use. Process changes which are to be considered in such case can be a major downside of it. Custom developed solutions are like doing own suite in a tailor shop. It will fit exactly your needs and requirements. Of course, that means that there will be considerable amount of time and financial resources invested.

Making a mistake in IT development company can also lead to dependency on one vendor who basically can be "racketeering" you. Technology stack which is offered in developing as well as customization projects is maybe complicated than ever before. More importantly than what does it cost is how long will it last and which further support you will be able to engage.

Implementation phase is having its own profile and can be profiled as beginning of the end. After beta testing period there's an option to go with "big bang" attitude of incremental start. If we decide to go fully with a new software starting on 1st of March, we are talking about big bang. No previous software will be used that way and full focus is on the new thing. Main issue in such approach can be that new software still can be having bugs or malfunctions what will reflect badly on implementation team. Also, skeptics will emphasize any issue as gigantic and using it as a proof that it would be better not to change anything. On the other side if everything goes well and education is prepared on time, there will be no need to come back on old solution.

Opposite on that approach, incremental process innovation considers step by step new software usage. It can be an optimal solution in cases when agile methodology provides sprints or modules by company department. Downside of this approach is an ability that employees procrastinate any new software usage since old one is available.

No matter on implementation steps, new software usage should be having an outstanding support. Again, if support fails to reply on client issues bad feedback will be given and resistance will grow. Comfortable and understanding support team can play a significant role in digital transformation process.

In literature you will find that digital transformation process is never ending story. It's good to empathize that such story has business competitiveness in focus so that is the main reason not to end. If all employees understand importance of fresh digital approach, new ideas will be coming and constant optimization together with new features will be an important factor of process improvement.

Not only software features (ex. during implementation important report was overseen) but also a business process improvement (ex. unnecessary scan is available to department for purchase) is expected in next iterations or software versions. Idea of continuous improvement within digital transformation is crucial if corporation is prepared for full and comprehensively process changes together with client needs.

It is also easy to conclude that digital transformation and digital growth is an ongoing process that is constantly evolving. Companies which are to be opened in wide technology specter – IoT, Machine learning, AI, Blockchain – together with "old" web 2.0 technology as well as strong focus on delivering value for customers – will assure its "place under the sun" in global market. AI can be used to automate processes and make decisions based on data, which can help organizations become more efficient and competitive. Blockchain can be used to manage digital records, such as contracts and financial transactions and provide higher levels of security and transparency. Finally, Total Quality Management (or its derivations) must be used to ensure that every step of a process is completed with quality in mind. By incorporating these technologies and mindsets as guidelines, organizations can take their digital transformation to the next level and remain competitive in the digital age.

SUSTAINABILITY AND AUTOMATION ARE ON THE RISE. REMOTE WORK AND BANDWIDTH ISSUES? NOT SO MUCH. THESE INSIGHTS SHOW WHAT'S HOT AND WHAT'S NOT IN 2023

Digital transformation is nothing new. Depending on how you define it, it can date back to the mid-twentieth century. Even by the most conservative interpretation, leading companies have been on the path to digitization for several decades.

Looking ahead to next year, it's clear that digital technologies will continue to play a critical role in corporate strategy and success. However, digital transformation has taken on a new urgency in the last three years. While companies have weathered the upheaval triggered by the pandemic, digitization has become an integral part of their responses and plans for the future.

However, certain aspects of digital transformation are likely to become more important while others will become less so. Below are some of the trends that IT expects to gain importance in 2022 – and others that are more likely to fade away.

Even companies that have embarked on a multi-year transformation journey have had to make mid-process adjustments, as they should. The operative word that defines the purpose of digital transformation in 2022 is resilience. The pandemic has taught companies that they must be prepared for seismic shifts in market dynamics and consumer needs. Forward-thinking companies will focus on effectively transitioning and managing change with minimal to no impact on internal and external customers.

Enterprises should start with their desired business outcome. Then CIOs should focus on restructuring, redesigning, and transforming their entire portfolio to achieve sustainable competitive advantage. The cloud is becoming the enabler, the delivery engine enabling this. All CIOs will prioritize cloud in the coming year because they recognize that cloud is more about innovation than cost savings.

"Artificial intelligence (AI) to drive greater process automation will continue to gain momentum as the technology matures and organizations are increasingly pressured to reduce costs through process automation," says Martha Heller, CEO of Heller Search Associates.

Some parts of the business have embraced workflow automation more eagerly than others. "For example, customer-facing teams like sales and marketing have typically used AI to connect disparate data and create intelligent workflows that enable team members to work smarter and more productively," says Christine Spang, co-founder, and CTO of Nylas. This is expected to spread throughout the company.

As we move into 2023 and beyond, we'll see workflow automation expand across the enterprise. That means HR and the HR teams will be able to schedule interviews and update candidate profiles faster and more efficiently, finance will update payment records automatically, and customer success teams will be better equipped to respond to customer needs quickly and efficiently.

"The remote effect has faded," says Jim Chilton, CIO at Cengage Group. Distributed work is no longer the exception but the rule. "We now know we can work from anywhere," Chilton says. "From a CIO or CTO perspective, that will be an integral part of our future.

Equipping and enabling remote and hybrid work remains a critical digital transformation priority through 2021. "While this will still be important in 2022, there has been enough investment in these areas over the past two years," says Everest Group's Joshi. "As a result, companies will reap the rewards of their investments in these areas rather than accelerate their focus."

THE FUTURE: MANAGING THE ENTIRE DATA LIFECYCLE

Increasingly, organizations and businesses are all about data. "Going forward, we need to rethink how we manage data from the cradle to the grave," says Melanie Kalmar, corporate vice president, CIO, and chief digital officer at Dow. "The technical organization plays a key role in shaping and leveraging data for the business to drive sustainability-driven decisions and reporting."

Interesting, yet challenging time to experience. What trends will still be with us after just three years and will there be the same Fortune 500 companies. Eager to see, stay tuned!

Kruno & Milan

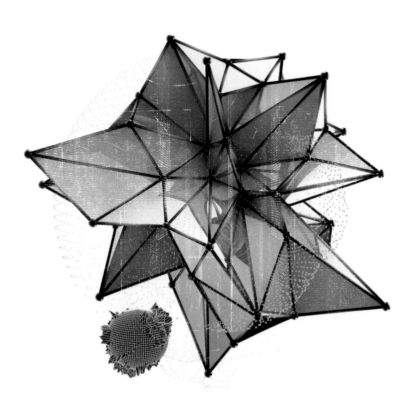

Websites

www.Accenture.com "Banking Consumer Study: Making digital more human"

www.acxiom.com "The Role of Data in Digital Transformation" https://www.acxiom.com/data-digital-transformation/

www.adobe.com "Defining digital transformation: Long-term evolution, not just a moment in time" https://blog.adobe.com/en/publish/2021/08/17/defining-digital-transformation-long-term-evolution-not-just-moment-in-time#gs.ps0xva

www.blog.panoply.io "The Data-Driven Society: Capturing the Value of Information" https://blog.panoply.io/the-data-driven-society-capturing-the-value-of-information

www.cardiff.ac.uk "What are the Implications of a Data-Driven Society?" https://www.cardiff.ac.uk/news/view/1186530-what-are-the-implications-of-a-data-driven-society

www.Cdc.gov: World Diabetes Day

www.cio.com "The 5 key drivers of digital transformation today" https://www.cio.com/article/230079/whats-now-in-digital-transformation.html

www.cmswire.com "Why Culture Change Is Essential for Digital Transformation" https://www.cmswire.com/digital-workplace/why-culture-change-is-essential-for-digital-transformation/

www.computerweekly.com "Digital transformation or digital evolution?" https://www.computerweekly.com/microscope/feature/Digital-transformation-or-digital-evolution

www.cordova.apache.org/

www.coredna.com "Digital Transformation Roadmap: 10 Steps to a Successful Digital Transformation" https://www.coredna.com/blogs/digital-transformation-roadmap#1

www.Deloitte.com: Wearable technology in health care: Getting better all the time

www.Digitalguardian.com: What is a Health Information System?

www.enterprisersproject.com "Digital transformation: 7 in-demand technology skills" https://enterprisersproject.com/article/2021/10/digital-transformation-7-skills-prioritize#:~:text=The%20core%20skills%20that%20organizations,The%20DevSecOps%20approach%20can%20help

www.europeanbusinessreview.com "The Role of Big Data in Digital Transformation" https://www.europeanbusinessreview.com/the-role-of-big-data-in-digital-transformation/ https://www.informationweek.com/big-data/the-importance-of-digital-transformation-in-predictive-analytics www.forbes.com "6 Predictions About The Future of Digital Transformation" https://www.forbes.com/sites/gilpress/2015/12/06/6-predictions-about-the-future-of-digital-transformation/?sh=2356fd101102

www.forbes.com "How the World Became Data-Driven, and What's Next" https://www.forbes.com/sites/googlecloud/2020/05/20/how-the-world-became-data-driven-and-whats-next/?sh=70e2b54257fc

www.forbes.com "The Evolution of Digital Transformation" https://www.forbes.com/sites/forbestechcouncil/2021/08/12/the-evolution-of-digital-transformation/?sh=1cdbeb806fb8

www.fromthesquare.org "The Limits of Knowledge in a Data-Driven Society" https://www.fromthesquare.org/the-limits-of-knowledge-in-a-data-driven-society/

www.game-learn.com "7 Examples of Successful Digital Transformation In Business" https://www.game-learn.com/en/resources/blog/7-examples-of-successful-digital-transformation-in-business/

www.gdsgrou.com " What is Digital Culture? Everything You Need To Know" https://gdsgroup.com/insights/it/what-is-digital-culture/#:~:text=A%20digital%20culture%20is%20a,think%20and%20communicate%20within%20society.&text=It's%20applicable%20to%20multiple%20topics,relationship%20between%20humans%20and%20technology

www.haptik.ai: "Elevated Customer Support for the World's Leading Insurance Company using AI."

www.hbr.org "Digital Transformation Comes Down to Talent in 4 Key Areas" https://hbr.org/2020/05/digital-transformation-comes-down-to-talent-in-4-key-areas

www.healthitsecurity.com: What Is Holding Healthcare Back from Digital Transformation

www.imd.org "Top 21 Digital Transformation Strategies" https://www.imd.org/imd-reflections/digital-programs-reflections/digital-transformation-strategies/

www.imnovation-hub.com "Big Data Analytics: the datafication of society" https://www.imnovation-hub.com/digital-transformation/big-data-analytics-the-datafication-of-society/

www.invonto.com "How to Develop a Winning Digital Transformation Strategy in 2021" https://www.invonto.com/insights/digital-transformation-strategy/

www.ionology.com "What is a Digital Transformation Strategy?" https://www.ionology.com/what-is-a-digital-transformation-strategy/

www.i-scoop.eu "Digital transformation: role and evolution of intelligent information" https://www.i-scoop.eu/digital-transformation/digital-transformation-intelligent-information-activation/

www.i-scoop.eu "What is digital business transformation? The essential guide to DX" https://www.i-scoop.eu/digital-transformation/

www.lumapps.com "10 Essential Tools to Support Your Digital Transformation" https://www.lumapps.com/solutions/digital-transformation/digital-transformation-tools/

www.mckinsey.com "Automating the insurance industry"

www.mckinsey.com: "How Insurance Can Prepare for the Next Distribution Model"

www.morethandigital.info "5 Trends As Drivers For The Digital Transformation of All Industries" https://morethandigital.info/en/5-trends-as-drivers-for-the-digital-transformation-of-all-industries/

www.One-beyond.com: Patient Benefits of Digital Transformation in Healthcare

www.poppulo.com "The 4 Main Areas of Digital Transformation" https://www.poppulo.com/blog/what-are-the-4-main-areas-of-digital-transformation

www.powerengieeringint.com "Digital transformation or digital evolution?" https://www.powerengineeringint.com/digitalization/digital-transformation-or-digital-evolution/

www.ptc.com "7 Tenents of an Effective Digital Transformation Strategy" https://www.ptc.com/en/blogs/corporate/digital-transformation-strategy#:~:text=What%20is%20a%20digital%20transformation,itself%2C%20a%20broad%20business%20strategy

www.superoffice.com "How Customer Experience Drives Digital Transformation" https://www.superoffice.com/blog/digital-transformation/

www.techgunnel.com "Why a Data-Driven Culture Is Critical to Digital Transformation" https://www.techfunnel.com/information-technology/data-driven-culture-to-accelerate-digital-transformation/

www.techrepublic.com "10 tech tools that will help bring digital transformation to your SMB" https://www.techrepublic.com/article/10-tech-tools-that-will-help-bring-digital-transformation-to-your-smb/

www.techrepublic.com "Digital Transformation Leads to Better Profits for 80% of Companies That Pursue it" https://www.techrepublic.com/article/digital-transformation-leads-to-better-profits-for-80-of-companies-that-pursue-it-says-report/

www.thepinnaclelist.com "Digital Transformation – Evolution or Revolution?" https://www.thepinnaclelist.com/articles/digital-transformation-evolution-revolution/#:~:text=Digital%20evolution%20simply%20means%20harnessing,you%20have%20built%20for%20years

www.U.plus: The State of Digital Transformation in Healthcare

www.virtu.com "8 Benefits of Digital Transformation" https://www.virtru.com/blog/8-benefits-digital-transformation/

www.wbresearch.com "Here's How Metromile Is Boosting Customer Experience with Digital Payments"

www.wgatfix.com "How to Deploy Your Digital Transformation Roadmap" https://whatfix.com/blog/digital-transformation-roadmap/

www.whatfix.com "What Is Digital Transformation?" https://whatfix.com/digital-transformation/

www.wilgroup.net "The Role of Agility in Digital Transformation" https://www.wilgroup.net/insights/the-role-of-agility-in-digital-transformation

www.zdnet.com "What is digital transformation? Everything you need to know about how technology is reshaping business" https://www.zdnet.com/article/what-is-digital-transformation-everything-you-need-to-know-about-how-technology-is-reshaping/

www2.deloitte.com "Start your digital transformation with the end in mind" https://www2.deloitte.com/us/en/insights/topics/digital-transformation/digital-transformation-evolution.html

Index

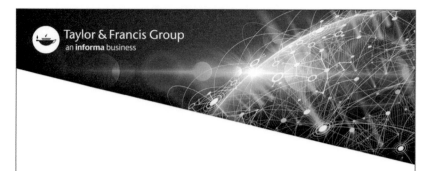